CYTOLOGY, GENETICS AND MOLECULAR BIOLOGY

CYTOLOGY, GENETICS AND MOLECULAR BIOLOGY

Shakib Ali
Md Salman Ahmed

Kruger Brentt
Publishers
2023

Kruger Brentt Publishers UK. LTD.
Company Number 9728962

Regd. Office: 68 St Margarets Road, Edgware, Middlesex HA8 9UU

ISBN: 9781787150638

For information on all our publications visit our website at http://krugerbrentt.com/

PREFACE

Cells are the fundamental unit of living things--they are the smallest structures that show all the features of living things. All organisms consist of small cells, typically too small to be seen by a naked eye, but big enough for an optical microscope. Each cell is a complex system consisting of many different building blocks enclosed in membrane bag. There are unicellular (consisting only of one cell) and multicellular organisms. Bacteria and baker's yeast are examples of unicellular organisms - any one cell is able to survive and multiply independently in appropriate environment.

After the generation of genome sequence data from a wide variety of plants, databases are filled with sequence information of genes with no known biological function, and while bioinformatics tools can help analyze genome sequences and predict gene structures, experimental approaches to discover gene functions need to be widely implemented. Molecular biology is the study of biology on molecular level. The field overlaps with areas of biology and chemistry, particularly genetics and biochemistry. Molecular biology chiefly concerns itself with understanding the interactions between the various systems of a cell, including the interactions between DNA (deoxyribonucleic acid), RNA (Ribonucleic acid) and protein biosynthesis as well as learning how these interactions are regulated

The present book contains twenty five chapters covering all related disciplines. These chapters include Cell theory, structure and function of cell, cytoplasm - structure and function, mendelism and chromosomal aberration, mutation, nucleus, nucleolus and chromosomes, intracellular compartments , the mitochondrion, cytoskeleton, genetics: an overview, mendels' laws of inheritance, lethality and interaction of genes, quantitative inheritance multiple alleles (based on classical concept of allelomorphism) physical basis of heredity (chromosome theory of heredity) linkage and crossing over sex linked, sex influenced and sex limited traits maternal effects, cytoplasmic inheritance and organellar genetics mutations: morphological level (including lethal mutations) mutations:

biochemical and microbial genetics structure and synthesis of nucleic acids structure and synthesis of proteins regulation of protein synthesis genetic code, overlaping genes and split genes recombinant dna, pcr and dna chips / microarrays synthesis, isolation and sequencing of genes. This book will provide a thorough understanding of the. Cytology, Genetics and Molecular Biology and provides the unique illustrations, providing students with strong molecular biology concepts. The book is also supported with a comprehensive glossary and index in the end. This text is designed for students taking a course in Molecular Biology, Cell Biology, Microbiology, Genetics, Biology, Pharmacology, Biotechnology, Biochemistry, and Agriculture.

We are grateful to all those persons as well as various books, manuals, periodicals, magazines, journals etc. that helped in the preparation of this book. In spite of the best efforts, it is possible that some errors may have occurred into the compilation and editing of the book. Further queries, constructive suggestions and criticisms for the improvement of the book are always welcome and shall be thankfully acknowledged.

Shakib Ali

Md Salman Ahmed

CONTENTS

1

CELL THEORY, STRUCTURE AND FUNCTION OF CELL

All organisms are composed of structural and functional units of life called 'cells'. The body of some organisms like bacteria, protozoans and some algae is made up of a single cell whereas the body of higher fungi, plants and animals are composed of many cells. Human body is built of about one trillion cells.

Cells vary in size and structure as they are specialized to perform different functions. But the basic components of the cell are common to all biological cells. This lesson deals with the structure common to all types of the cells. You will also learn about the kinds of cell division and the processes involved therein in this lesson.

Cell biology is a branch of biology, which is mainly involved with the study of the cell, its types, structure, functions and its interaction with other cells. Cell biology is also called cytology and is derived from the Greek word "kytos", meaning vessel.

IMPORTANT TOPICS IN CELL BIOLOGY

Cell

Cells are the basic, fundamental unit of life. So, if we were to break apart an organism to the cellular level, the smallest independent component that we would find would be the cell.

Explore the cell notes to know what is a cell, cell definition, cell structure, types and functions of cells. These notes have an in-depth description of all the concepts related to cells.

Cell Definition

"A cell is defined as the smallest, basic unit of life that is responsible for all of life's processes."

Cells are the structural, functional, and biological units of all living beings. A cell can replicate itself independently. Hence, they are known as the building blocks of life.

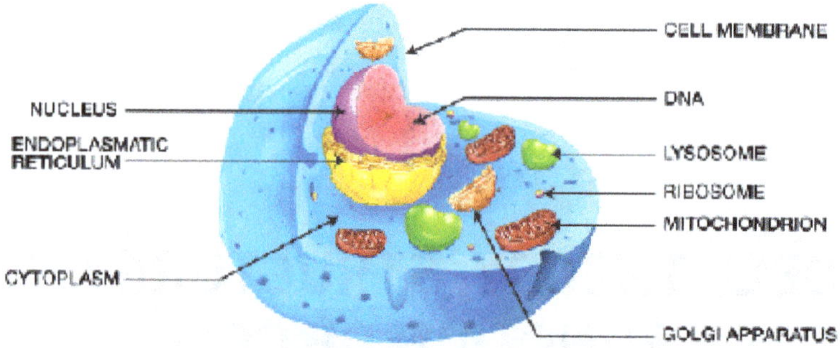

Fig. 1 Cells are the fundamental unit of life. They range in size from 0.0001 mm to nearly 150 mm across

Each cell contains a fluid called the cytoplasm, which is enclosed by a membrane. Also present in the cytoplasm are several biomolecules like proteins, nucleic acids and lipids. Moreover, cellular structures called cell organelles are suspended in the cytoplasm.

What is a Cell?

A cell is the structural and fundamental unit of life. The study of cells from its basic structure to the functions of every cell organelle is called Cell Biology. Robert Hooke was the first Biologist who discovered cells.

All organisms are made up of cells. They may be made up of a single cell (unicellular), or many cells (multicellular). Mycoplasmas are the smallest known cells. Cells are the building blocks of all living beings. They provide structure to the body and convert the nutrients taken from the food into energy.

Cells are complex and their components perform various functions in an organism. They are of different shapes and sizes, pretty much like bricks of the buildings. Our body is made up of cells of different shapes and sizes.

Cells are the lowest level of organisation in every life form. From organism to organism, the count of cells may vary. Humans have the number of cells compared to that of bacteria.

Cells comprise several cell organelles that perform specialised functions to carry out life processes. Every organelle has a specific structure. The hereditary material of the organisms is also present in the cells.

Discovery of Cells

Discovery of cells is one of the remarkable advancements in the field of science. It helps us know that all the organisms are made up of cells, and these cells help in carrying out various life processes. The structure and functions of cells helped us to understand life in a better way.

Robert Hooke discovered the cell in 1665. Robert Hooke observed a piece of bottle cork under a compound microscope and noticed minuscule structures that reminded him of small rooms. Consequently, he named these "rooms" as cells. However, his compound microscope had limited magnification, and hence, he could not see any details in the structure. Owing to this limitation, Hooke concluded that these were non-living entities.

Later Anton Van Leeuwenhoek observed cells under another compound microscope with higher magnification. This time, he had noted that the cells exhibited some form of movement (motility). As a result, Leeuwenhoek concluded that these microscopic entities were "alive." Eventually, after a host of other observations, these entities were named as animalcules.

In 1883, Robert Brown, a Scottish botanist, provided the very first insights into the cell structure. He was able to describe the nucleus present in the cells of orchids.

Characteristics of Cells

Following are the various essential characteristics of cells:

- Cells provide structure and support to the body of an organism.
- The cell interior is organised into different individual organelles surrounded by a separate membrane.
- The nucleus (major organelle) holds genetic information necessary for reproduction and cell growth.
- Every cell has one nucleus and membrane-bound organelles in the cytoplasm.
- Mitochondria, a double membrane-bound organelle is mainly responsible for the energy transactions vital for the survival of the cell.
- Lysosomes digest unwanted materials in the cell.
- Endoplasmic reticulum plays a significant role in the internal organisation of the cell by synthesising selective molecules and processing, directing and sorting them to their appropriate locations.

TYPES OF CELLS

Cells are similar to factories with different labourers and departments that work towards a common objective. Various types of cells perform different functions. Based on cellular structure, there are two types of cells:

- Prokaryotes
- Eukaryotes

Prokaryotic Cells

1. Prokaryotic cells have no nucleus. Instead, some prokaryotes such as bacteria have a region within the cell where the genetic material is freely suspended.

This region is called the nucleoid.

2. They all are single-celled microorganisms. Examples include archaea, bacteria, and cyanobacteria.
3. The cell size ranges from 0.1 to 0.5 μm in diameter.
4. The hereditary material can either be DNA or RNA.
5. Prokaryotes generally reproduce by binary fission, a form of asexual reproduction. They are also known to use conjugation – which is often seen as the prokaryotic equivalent to sexual reproduction (however, it is NOT sexual reproduction).

Eukaryotic Cells

1. Eukaryotic cells are characterised by a true nucleus.
2. The size of the cells ranges between 10–100 μm in diameter.
3. This broad category involves plants, fungi, protozoans, and animals.
4. The plasma membrane is responsible for monitoring the transport of nutrients and electrolytes in and out of the cells. It is also responsible for cell to cell communication.
5. They reproduce sexually as well as asexually.
6. There are some contrasting features between plant and animal cells. For eg., the plant cell contains chloroplast, central vacuoles, and other plastids, whereas the animal cells do not.

CELL STRUCTURE

The cell structure comprises individual components with specific functions essential to carry out life's processes. These components include- cell wall, cell membrane, cytoplasm, nucleus, and cell organelles. Read on to explore more insights on cell structure and function.

Cell Membrane

- The cell membrane supports and protects the cell. It controls the movement of substances in and out of the cells. It separates the cell from the external environment. The cell membrane is present in all the cells.
- The cell membrane is the outer covering of a cell within which all other organelles, such as the cytoplasm and nucleus, are enclosed. It is also referred to as the plasma membrane.
- By structure, it is a porous membrane (with pores) which permit the movement of selective substances in and out of the cell. Besides this, the cell membrane also protects the cellular component from damage and leakage.
- It forms the wall-like structure between two cells as well as between the cell and its surroundings.

- Plants are immobile, so their cell structures are well-adapted to protect them from external factors. The cell wall helps to reinforce this function.

CELL WALL

- The cell wall is the most prominent part of the plant's cell structure. It is made up of cellulose, hemicellulose and pectin.
- The cell wall is present exclusively in plant cells. It protects the plasma membrane and other cellular components. The cell wall is also the outermost layer of plant cells.
- It is a rigid and stiff structure surrounding the cell membrane.
- It provides shape and support to the cells and protects them from mechanical shocks and injuries.

Cytoplasm

- The cytoplasm is a thick, clear, jelly-like substance present inside the cell membrane.
- Most of the chemical reactions within a cell take place in this cytoplasm.
- The cell organelles such as endoplasmic reticulum, vacuoles, mitochondria, ribosomes, are suspended in this cytoplasm.

Nucleus

- The nucleus contains the hereditary material of the cell, the DNA.
- It sends signals to the cells to grow, mature, divide and die.
- The nucleus is surrounded by the nuclear envelope that separates the DNA from the rest of the cell.
- The nucleus protects the DNA and is an integral component of a plant's cell structure.

CELL ORGANELLES

Cells are composed of various cell organelles that perform certain specific functions to carry out life's processes. The different cell organelles, along with its principal functions, are as follows:

Cell Organelle and its Functions

Nucleolus

The nucleolus is the site of ribosome synthesis. Also, it is involved in controlling cellular activities and cellular reproduction

Nuclear membrane

The nuclear membrane protects the nucleus by forming a boundary between the nucleus and other cell organelles.

Chromosomes

Chromosomes play a crucial role in determining the sex of an individual. Each human cells contain 23 pairs of chromosomes

Endoplasmic reticulum

The endoplasmic reticulum is involved in the transportation of substances throughout the cell. It plays a primary role in the metabolism of carbohydrates, synthesis of lipids, steroids and proteins.

Golgi Bodies

Golgi bodies are called the cell's post office as it is involved in the transportation of materials within the cell

Ribosome

Ribosomes are the protein synthesisers of the cell

Mitochondria

The mitochondrion is called "the powerhouse of the cell." It is called so because it produces ATP – the cell's energy currency

Lysosomes

Lysosomes protect the cell by engulfing the foreign bodies entering the cell and helps in cell renewal. Therefore, it is known as the cell's suicide bags

Chloroplast

Chloroplasts are the primary organelles for photosynthesis. It contains the pigment chlorophyll

Vacuoles

Vacuoles stores food, water, and other waste materials in the cell

CELL THEORY

Cell Theory was proposed by the German scientists, Theodor Schwann, Matthias Schleiden, and Rudolf Virchow. The cell theory states that:

- All living species on Earth are composed of cells.
- A cell is the basic unit of life.
- All cells arise from pre-existing cells.

A modern version of the cell theory was eventually formulated, and it contains the following postulates:

- Energy flows within the cells.
- Genetic information is passed on from one cell to the other.
- The chemical composition of all the cells is the same.

Functions of Cell

A cell performs these major functions essential for the growth and development of an organism. Important functions of cell are as follows:

Provides Support and Structure

All the organisms are made up of cells. They form the structural basis of all the organisms. The cell wall and the cell membrane are the main components that function to provide support and structure to the organism. For eg., the skin is made up of a large number of cells. Xylem present in the vascular plants is made of cells that provide structural support to the plants.

Facilitate Growth Mitosis

In the process of mitosis, the parent cell divides into the daughter cells. Thus, the cells multiply and facilitate the growth in an organism.

Allows Transport of Substances

Various nutrients are imported by the cells to carry out various chemical processes going on inside the cells. The waste produced by the chemical processes is eliminated from the cells by active and passive transport. Small molecules such as oxygen, carbon dioxide, and ethanol diffuse across the cell membrane along the concentration gradient. This is known as passive transport. The larger molecules diffuse across the cell membrane through active transport where the cells require a lot of energy to transport the substances.

Energy Production

Cells require energy to carry out various chemical processes. This energy is produced by the cells through a process called photosynthesis in plants and respiration in animals.

Aids in Reproduction

A cell aids in reproduction through the processes called mitosis and meiosis. Mitosis is termed as the asexual reproduction where the parent cell divides to form daughter cells. Meiosis causes the daughter cells to be genetically different from the parent cells. Thus, we can understand why cells are known as the structural and functional unit of life. This is because they are responsible for providing structure to the organisms and performs several functions necessary for carrying out life's processes.

CELL DIVISION

Cell division happens when a parent cell divides into two or more cells called daughter cells. Cell division usually occurs as part of a larger cell cycle. All cells reproduce by splitting into two, where each parental cell gives rise to two daughter cells.

In other words, such cycles of growth and division allow a single cell to form a structure consisting of millions of cells.

Explore the cell division notes to learn about the types and phases of cell division. These newly formed daughter cells could themselves divide and grow, giving rise to a new cell population that is formed by the division and growth of a single parental cell and its descendant.

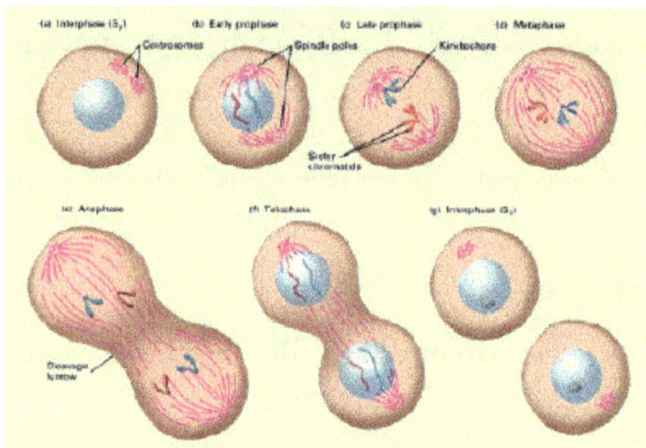

Fig. 2 Cell Division

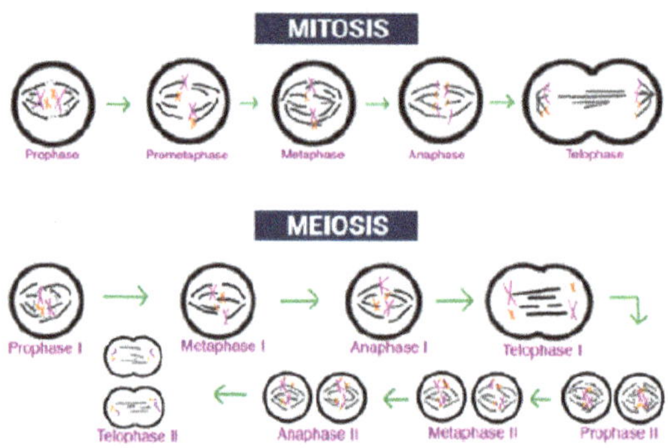

Fig. 3. Types of Cell Division

Types of Cell Division

There are two distinct types of cell division out of which the first one is vegetative division, wherein each daughter cell duplicates the parent cell called mitosis. The second one is meiosis, which divides into four haploid daughter cells.

Mitosis: The process cells use to make exact replicas of themselves. Mitosis is observed in almost all the body's cells, including eyes, skin, hair, and muscle cells.

Meiosis: In this type of cell division, sperm or egg cells are produced instead of identical daughter cells as in mitosis.

Binary Fission: Single-celled organisms like bacteria replicate themselves for reproduction.

Phases of the Cell Cycle

There are two primary phases in the cell cycle:

1. **Interphase:** This phase was thought to represent the resting stage between subsequent cell divisions, but new research has shown that it is a very active phase.
2. **M Phase (Mitosis phase):** This is where the actual cell division occurs. There are two key steps in this phase, namely cytokinesis and karyokinesis.
3. The interphase further comprises three phases:
4. **G0 Phase (Resting Phase):** The cell neither divides nor prepares itself for the division.
5. **G1 Phase (Gap 1):** The cell is metabolically active and grows continuously during this phase.
6. **S phase (Synthesis):** The DNA replication or synthesis occurs during this stage.
7. **G2 phase (Gap 2):** Protein synthesis happens in this phase.
8. **Quiescent Stage (G0):** The cells that do not undergo further division exits the G1 phase and enters an inactive stage. This stage is known as the quiescent stage (G0) of the cell cycle.

There are four stages in the M Phase, namely:

1. Prophase
2. Metaphase
3. Anaphase
4. Telophase

ANIMAL CELL

We are aware of the fact that the cell is the structural and fundamental unit of life. It is also the smallest and the most basic biological unit of living organisms. On the basis of the cellular organization, cells are further classified as eukaryotic and prokaryotic. Plant cells and animal cells fall under the eukaryotic category.

Animal Cell Definition

"An animal cell is a type of eukaryotic cell that lacks a cell wall and has a true, membrane-bound nucleus along with other cellular organelles."

Explanation

Animal cells range in size from a few microscopic microns to few millimetres. The largest known animal cell is the ostrich egg, which can stretch over 5.1 inches across and weighs about 1.4 kilograms. This is in stark contrast to the neuron in the human body, which is just 100 microns across.

The shape of animal cells also varies, with some being flat, others oval or rod-shaped. There are also more intriguing shapes such as curved, spherical, concave and rectangular. Most of the cells are microscopic in size and can only be seen under the microscope.

As stated before, animal cells are eukaryotic cells with a membrane-bound nucleus. Furthermore, these cells exhibit the presence of DNA inside the nucleus. They also comprise other membrane-bound organelles and cellular structures which carry out specific functions necessary for a cell to function properly.

Animal Cell Diagram

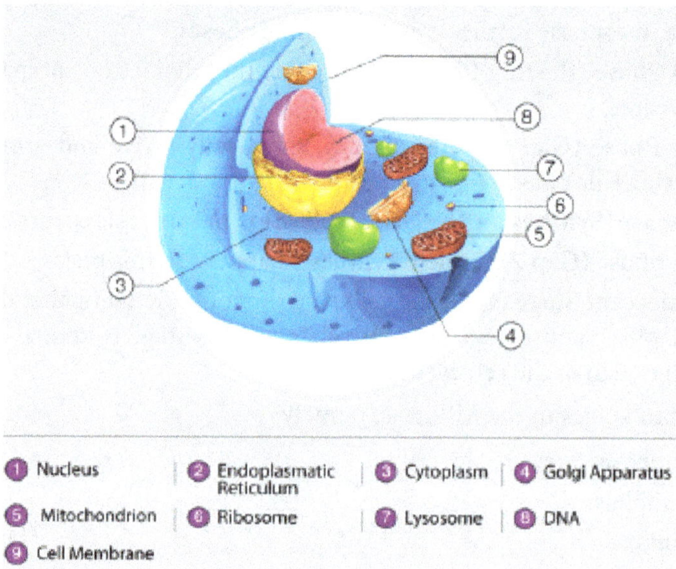

Fig. 4 Animal cell diagram detailing the various organelles

The diagram given below depicts the structural organization of the animal cell. The various cell organelles present in an animal cell are clearly marked in the animal cell diagram provided below.

Though this animal cell diagram is not representative of any one particular type of cell, it provides insight into the primary organelles and the intricate internal structure of most animal cells. Furthermore, it is easy to distinguish between a plant and animal cell diagram just by inspecting the presence or absence of a cell wall.

Animal Cell Structure

Animal cells are generally smaller than plant cells. Another defining characteristic is its irregular shape. This is due to the absence of a cell wall. But animal cells share other cellular organelles with plant cells as both have evolved from eukaryotic cells.

A typical animal cell comprises the following cell organelles:

Cell Membrane

A thin semipermeable membrane layer of protein and fats surrounding the cell. Its primary role is to protect the cell from its surrounding. Also, it controls the entry and exit of nutrients and other microscopic entities into the cell.

Nuclear Membrane

It is a double-membrane structure that surrounds the nucleus. It is also referred to as the nuclear envelope.

Nucleus

It is an organelle that contains several other sub-organelles such as nucleolus, nucleosomes and chromatins. It also contains DNA and other genetic materials.

Centrosome

It is a small organelle found near to the nucleus which has a thick centre with radiating tubules. The centrosomes are where microtubules are produced.

Lysosome (Cell Vesicles)

They are round organelles surrounded by a membrane and comprising digestive enzymes which help in digestion, excretion and in the cell renewal process.

Cytoplasm

A jelly-like material which contains all the cell organelles, enclosed within the cell membrane. The substance found within the cell nucleus, contained by the nuclear membrane is called the nucleoplasm.

Golgi Apparatus

A flat, smooth layered, sac-like organelle which is located near the nucleus and involved in manufacturing, storing, packing and transporting the particles throughout the cell.

Mitochondrion

They are spherical or rod-shaped organelles with a double membrane. They are the powerhouse of a cell as they play an important role in releasing energy.

Ribosome

They are small organelles made up of RNA-rich cytoplasmic granules, and they are the sites of protein synthesis.

Endoplasmic Reticulum (ER)

This cellular organelle is composed of a thin, winding network of membranous sacs originating from the nucleus.

Vacuole

A membrane-bound organelle present inside a cell involved in maintaining shape and storing water, food, wastes, etc.

Nucleopore

They are tiny holes present in the nuclear membrane which are involved in the movement of nucleic acids and proteins within the cell.

Animal Cell Types

There are numerous types of animal cells, each designed to serve specific functions. The most common types of animal cells are:

Skin Cells

Melanocytes, keratinocytes, Merkel cells and Langerhans cells

Muscle Cells

Myocyte, Myosatellite cells, Tendon cells, Cardiac muscle cells

Blood Cells

Leukocytes, erythrocytes, platelet

Nerve Cells

Schwann cell, glial cells etc

Fat Cells

Adipocytes

Points to Note About Animal Cell

The cell is the structural and functional unit of life. These cells differ in their shapes, sizes and their structure as they have to fulfil specific functions. Plant cells and animal cells share some common features as both are eukaryotic cells.

However, they differ as animals need to adapt to a more active and non-sedentary lifestyle. Furthermore, animals need to acquire their own food, therefore, they do not possess any of the specialized cell organelles such as chloroplasts.

PLANT CELLS - DEFINITION, DIAGRAM, STRUCTURE & FUNCTION

The cell is the basic unit of life in all organisms. Like humans and animals, plants are also composed of several cells. The plant cell is surrounded by a cell wall which is involved in providing shape to the plant cell. Apart from the cell wall, there are other organelles that are associated with different cellular activities.

Let us have a detailed look at the plant cell, its structure and functions of different plant cell organelles.

Plant Cell Definition

"Plant cells are eukaryotic cells with a true nucleus along with specialized structures called organelles that carry out certain specific functions."

WHAT IS A PLANT CELL?

Plant cells are eukaryotic cells that vary in several fundamental factors from other eukaryotic organisms. Both plant and animal cells contain nucleus along with similar organelles. One of the distinctive aspects of a plant cell is the presence of a cell wall outside the cell membrane.

Plant Cell Diagram

The plant cell is rectangular and comparatively larger than the animal cell. Even though plant and animal cells are eukaryotic and share a few cell organelles, plant cells are quite distinct when compared to animal cells as they perform different functions. Some of these differences can be clearly understood when the cells are examined under an electron microscope.

Plant Cell Structure

Just like different organs within the body, plant cell structure includes various components known as cell organelles that perform different functions to sustain itself. These organelles include:

Cell Wall

It is a rigid layer which is composed of cellulose, glycoproteins, lignin, pectin and hemicellulose. It is located outside the cell membrane. It comprises proteins, polysaccharides and cellulose.

The primary function of the cell wall is to protect and provide structural support to the cell. The plant cell wall is also involved in protecting the cell against mechanical stress and to provide form and structure to the cell. It also filters the molecules passing in and out of the cell.

The formation of the cell wall is guided by microtubules. It consists of three layers, namely, primary, secondary and the middle lamella. The primary cell wall is formed by cellulose laid down by enzymes.

Cell membrane

It is the semi-permeable membrane that is present within the cell wall. It is composed of a thin layer of protein and fat.

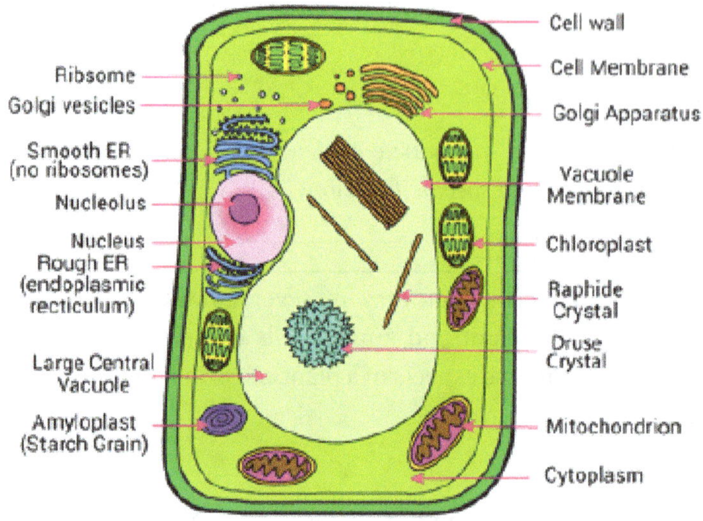

Fig. 5 Plant Cell Diagram showing different cell organelles

The cell membrane plays an important role in regulating the entry and exit of specific substances within the cell.

For instance, cell membrane keeps toxins from entering inside, while nutrients and essential minerals are transported across.

CELL WALL

There are many forms of life on earth with a multitude of features and characteristics. Yet, when we look at the microscopic level, every single life on earth is composed of the same basic, structural and functional unit of life – the cell.

The types of cell, chemical composition, and its characteristics vary depending on the type of life. One of those important characteristics is the cell wall. Let us explore more about the cell wall, its types and where it is found.

What is a Cell Wall?

A cell wall is defined as the non-living component, covering the outmost layer of a cell. Its composition varies according to the organism and is permeable in nature. The cell wall separates the interior contents of the cell from the exterior environment. It also provides shape, support, and protection to the cell and its organelles. However, this cellular component is present exclusively in eukaryotic plants, fungi, and few prokaryotic organisms.

As stated above, fungi also possess cell walls, but they are made up of chitin, a derivative of glucose which is also found in the exoskeletons of arthropods. And just like the cell walls in plants, they provide structural support and prevents desiccation.

Prokaryotic organisms such as bacteria also contain cell walls. However, they are chemically different from the cell wall found in plants and fungi. The prokaryotic cell walls are composed of large polymers known as peptidoglycans. Cell walls in prokaryotes serve as a form of protection and prevent lysis (bursting of the cell and expulsion of cellular contents). Structurally, prokaryotic cell walls consist of two layers:

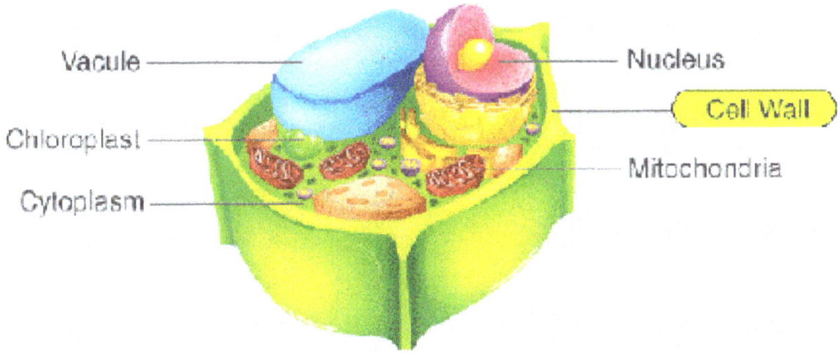

Fig. 6 A cross-section of a plant cell with the cell wall highlighted

1. An inner layer that is made up of peptidoglycans
2. An outer layer that is composed of lipoproteins and lipopolysaccharides

Eukaryotic cells possess a definite nucleus along with a distinct nuclear membrane. It also contains membrane-bound organelles not found in prokaryotic cells. Another important point to note is that the cell wall is absent in other eukaryotic organisms such as animals, only plants possess cell walls.

Cell Wall Structure

The cell wall is the outer covering of a cell, present adjacent to the cell membrane, which is also called the plasma membrane. As mentioned earlier, the cell wall is present in all plant cells, fungi, bacteria, algae, and some archaea. An animal cell is irregular in their shape and this is mainly due to the lack of cell wall in their cells. The compositions of the cell wall usually vary along with organisms.

The plant cell wall is generally arranged in three layers and composed of carbohydrates, like pectin, cellulose, hemicellulose and other smaller amounts of minerals, which form a network along with structural proteins to form the cell wall. The three major layers are:

1. Primary Cell Wall
2. The Middle Lamella
3. The Secondary Cell Wall

Primary Cell Wall

The primary cell is situated closest to the inside of the cell and is the first-formed cell wall. It is mainly made up of cellulose, allowing the wall to stretch for the purpose of growth. Several primary cells contain pectic polysaccharides and structural proteins. It is also comparatively permeable and thinner than the other layers.

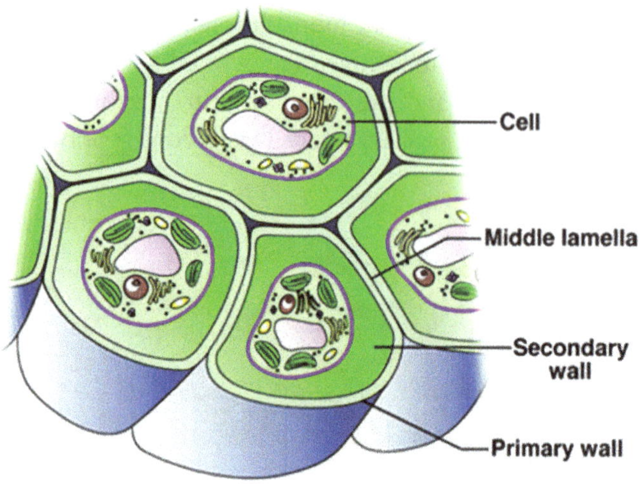

Fig. 7 A plant cell showcasing the primary cell wall, middle lamella, and the secondary cell wall

Middle Lamella

The middle lamella is also the outermost layer and it acts as an interface between the other neighbouring cells and glues them together. This layer primarily consists of pectins. However, other substances such as lignin and proteins can also be found.

Secondary Cell Wall

The secondary cell wall is formed inside the primary cell wall once the cell is completely grown. Some types of cells (especially the cells of xylem tissues) consist of cellulose and lignin and these provide additional rigidity and waterproofing. Also, this layer provides the characteristic rectangular or square shape to a cell. It is also the thickest layer and permits permeability.

FUNCTION OF THE CELL WALL

The cell wall is an integral component of the plant cell and it performs many essential functions. Following are some of the major cell wall functions observed:

- The plant cell wall provides definite shape, strength, rigidity
- It also provides protection against mechanical stress and physical shocks
- It helps to control cell expansion due to the intake of water
- Also helps in preventing water loss from the cell
- It is responsible for transporting substances between and across the cell
- It acts as a barrier between the interior cellular components and the external environment

CELL WALL AND CELL MEMBRANE

A cell is the structural and functional working unit of life and also been described as building blocks and fundamental unit of an organism. The term cell was coined by an English scientist Robert Hooke in the year 1665. The shape and size of the cells vary according to their functions and compositions. There are different types of cells and can be differentiated based on the presence and absence of few cell organelles.

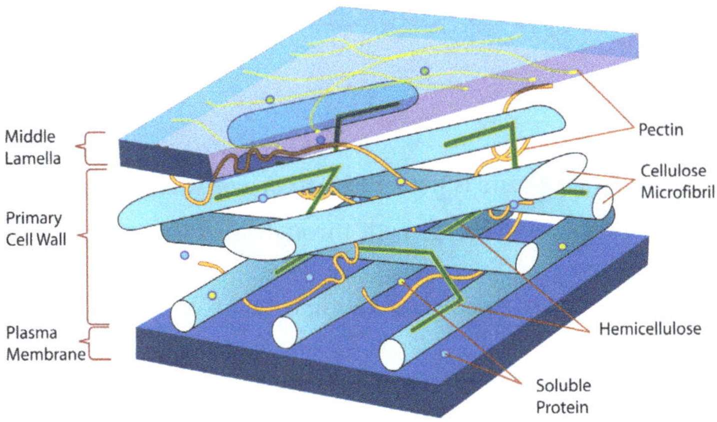

Fig. 8 Tthe various components of a plant cell wall

The cell wall is present only in plants, fungi and bacteria. The image above represents a plant cell wall

The cell wall is the outermost covering of plant cells. It is present outside the cell membrane and is tough, flexible and sometimes rigid in its texture. It is mainly composed of cellulose, long fibres of carbohydrates including hemicellulose, lignin and pectin.

The main functions of the cell wall are:

1. Protecting the cell against physical damage and invading pathogens.
2. Regulates and controls the direction of cell growth.
3. Providing the strength, structural support and maintaining the shape of the cell.
4. Functions as a storage unit by storing carbohydrates for use in plant growth, especially in seeds.
5. It allows entry of smaller molecules through it freely.

The cell membrane is also known as the plasma membrane. It is the outermost covering of animal cells. It is a semi-permeable membrane composed of lipids and proteins. The main functions of the cell membrane include:

1. Protecting the integrity of the interior cell.
2. Providing support and maintaining the shape of the cell.

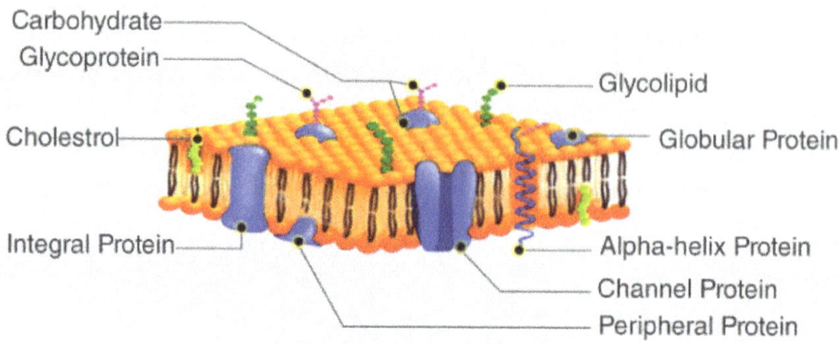

Fig. 9 Cell Membrane is present in all organisms including plants

3. Helps in regulating cell growth through the balance of endocytosis and exocytosis.
4. The cell membrane also plays an important role in cell signalling and communication.
5. It acts as a selectively permeable membrane by allowing the entry of only selected substances into the cell.

Difference between Cell Wall and Cell Membrane

One of the fundamental differences between the plasma membrane and the cell wall is in the type of organisms that they are found. The cell wall is present only in plants and the cell membrane is present in every living organism including plants.

CELL WALL	CELL MEMBRANE
Present only in plants and in some fungi, bacteria, algae.	Present in all types of cells, in humans, animals, plants, bacteria, etc.
It is the outermost part of the plant cell	It is the outermost covering the animal cells
It is made up of pectin, chitin, lignin, glycoproteins, glycolipids, sugar, and cellulose.	It is a lipid bilayer. And is composed of lipoproteins and carbohydrates.
The cell wall is 0.1 μm to several μm in thickness	The cell membrane is 7.5–10 nm in thickness
It is the thick and rigid structure with a fixed shape.	It is a thin and delicate structure. It is flexible to change the shape as needed.
It protects the cell from the external environment	It protects and maintains the internal environment of the cell.
The cell wall is metabolically inactive	The cell membrane is metabolically active.
The cell wall grows in thickness over time.	It is of the same thickness for the lifetime of the organism.
Further, it occupies the whole cell in the plant as the cell ages and dies.	

CELL WALL	CELL MEMBRANE
The cell wall is fully permeable to smaller molecules with the size of 30-60 kDa.	The membrane is selectively permeable and controls the movement of the substance into and outside the cell.
Functions include protection from the external environment.	Functions include permeability, signal reception, motility conduction, cell division, sexual reproduction, etc.

CONCLUSION

The cell wall and cell membrane are two organelles that are an important part of living organisms. However, the former is found only in plants, fungi and in some bacteria. Traditionally, a cell well is defined as the layer of polysaccharides that exists outside the plasma membrane. It is rigid and serves structural and supportive functions. On the other hand, the cell wall in fungi is made from another material called chitin, which is also found in the exoskeletons of arthropods.

The cell membrane is present in all living organisms, including plants. It is the outermost layer of the cell in animals and encloses other cellular organelles within. Unlike the cell wall, the cell membrane is flexible and its shape can be changed as needed. Also, it is metabolically active and has selective permeability.

2

CYTOPLASM - STRUCTURE AND FUNCTION

BRIEF HISTORY OF CYTOPLASM

In the year 1665, Robert Hooke, an English researcher observed the fundamental unit of life through his coarse compound microscope. He coined the term "cell", which was based on the Latin word called "Cella," meaning, small rooms.

Later, many scientists contributed to Robert Hooke's findings and eventually, the Cell Theory was postulated. As technology progressed, modern interpretations of Cell Theory have been formed, with new tenets and revisions of the existing ones.

In 1863, a Swiss biologist named Rudolf von Kölliker coined the term "Cytoplasm," but it was regarded as a synonym for protoplasm. However, the term gradually changed its meaning to the current definition of the term "cytoplasm".

INTRODUCTION

The fluid that fills up the cells is referred to as the cytoplasm. It encompasses the cytosol with filaments, ions, proteins, and macromolecular structures and also other organelles suspended in the cytosol.

But new research suggests that the traditional definition of cytoplasm is no longer valid. Decades prior, it was considered to be a fluid-like substance, but new evidence reveals that it is similar to glass-forming liquids.

The cytoplasm in the eukaryotic cells associates with the cell contents except for the nucleus. But in prokaryotic cells, as they do not possess a defined nuclear membrane, the cytoplasm possesses the genetic material of the cell. The cells, in comparison to the eukaryotes, are smaller and have an uncomplicated arrangement of the cytoplasm.

Cytoplasm Structure

Cell organelles are various structures existing inside cells. All these structures are distinct and perform specific functions. Cells have three main elements i.e., plasma membrane, and cytoplasm and the nucleus.

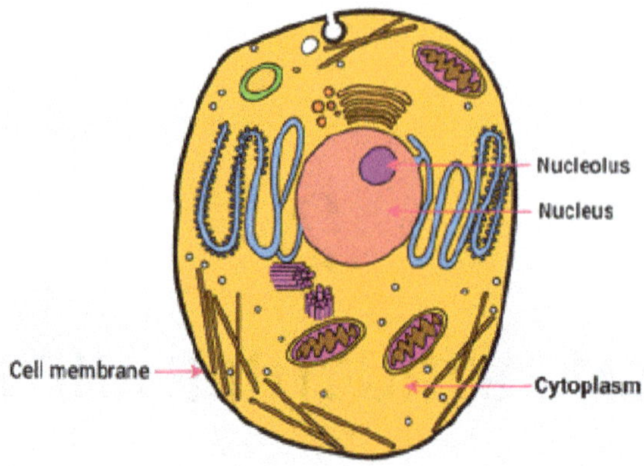

Fig. 1 The cytoplasm is a fluid-like substance that fills up the cells. Cellular organelles and structures are suspended in the cytoplasm.

The plasma membrane or cell membrane is a bi-lipid membranous layer, parting the cell organelles from its outside environment and from the different cells. It is the external covering of a cell where all different parts, including cytoplasm and nucleus, are enclosed.

Next, is the nucleus, one of the biggest organelle. They have exclusive control of a cell. Lastly, the cytoplasm is a jelly-like material in which the cell organelles are implanted.

The cytoplasm is an essential component of the cell. It is a semi-liquid jelly-like material, which joins the nucleus and the cell membrane. In the cell, the cytoplasm is embedded, while other cell organelles such as endoplasmic reticulum, mitochondria, ribosomes, vacuoles, etc. are all suspended within it.

It can easily be examined under a microscope through the staining technique. Functionally, it is the site for several chemical reactions within a cell. Most of the cellular metabolism takes place here.

CYTOPLASM FUNCTION

One of the major functions of cytoplasm is to enable cells to maintain their turgidity, which enables the cells to hold their shape. Other functions of cytoplasm are as follows:

- The jelly-like fluid of the cytoplasm is composed of salt and water and is present within the membrane of the cells and embeds all of the parts of the cells and organelles.
- The cytoplasm is home to many activities of the cell as it contains molecules, enzymes that are crucial in the break down of the waste.
- The cytoplasm also assists in metabolic activities.
- Cytoplasm provides shape to the cell. It fills up the cells thus enabling the organelles to remain in their position. The cells, without cytoplasm, would deflate and substances will not permeate easily from one to the other organelle.

- A part of the cytoplasm, the cytosol has no organelles. Rather, the cytosol is enclosed by matrix boundaries that fill up the cell section which does not hold the organelles.

The whole cellular content of a living cell is called protoplasm. The cytoplasm, nucleus and all other living components of the cell together make up the protoplasm of a cell.

Protoplasm

The protoplasm is generally referred to as the living part of the cell. It is the colourless, jelly-like substance composed of macromolecules, water and mixture of small molecules. It can be defined as the inorganic and organic substance which constitutes the cytoplasm, the nucleus, mitochondria and the plastids of the cell. It is the chief substance that is responsible for all the living processes.

The primary component of the protoplasm is the cytoplasm which is situated between the nucleus and the cell membrane in the eukaryotic cells. It contains all the organelles. It regulates the environment of the cell and maintains the cell shape. It stores the substances and chemicals that are necessary for the organelle.

The second component of the protoplasm is the nucleus, which contains the genetic material of an organism and is situated in the nucleus. The nucleus contains the ribosomes that are required and are crucial for the protein synthesis in the cells.

But in the case of prokaryotes, the nucleoid is present in place of the nucleus, wherein all the genetic information is present. However, it does not have a nuclear membrane; hence, the term protoplasm does not apply.

The elements that make up the protoplasm are fats, proteins, enzymes, hormones, etc. which are either suspended or dissolved in the fluid component of the protoplasm.

Nucleus

The nucleus is an important element of the cells. It is a membrane-bound organelle that is typically found in the eukaryotic cells. Eukaryotes usually have a single nucleus, whereas some cell types do not possess a nucleus (RBC-Red Blood Cells).

Cytoskeleton

"Cytoskeleton is the structure that maintains the shape and internal organization of the cell, and provides it mechanical support. "

The cytoskeleton is the network of fibres forming the eukaryotic cells, prokaryotic cells and archaeans. These fibres in the eukaryotic cells contain a complex mesh of protein filaments and motor proteins that help in cell movement.

It provides shape and support to the cell, organizes the organelles and facilitates transport of molecules, cell division and cell signalling.

Cytoskeleton Structure

A cytoskeleton structure comprises the following types of fibres:

- Microfilaments
- Microtubules
- Intermediate Filaments

Microtubules

Microtubules appear like small, hollow, round tubes with a diameter of about 24 nanometers. They are made up of a protein, tubulin. Thirteen tubulins link to form a single tube. Microtubules are very dynamic structures, which reveal that they can change quickly. They keep growing or shrinking steadily. These help in transporting cellular materials and dividing chromosomes during cell division.

Microfilaments

Microfilaments are thread-like protein fibres, 3-6 nm in diameter. They are particularly found in muscle cells. They consist of the protein actin, responsible for muscle contraction. These are also responsible for cellular movements including cytokinesis, contraction, and gliding.

Intermediate Filaments

The intermediate filaments are about 10 nm in diameter and provide tensile strength to the cell. They facilitate the formation of keratins and neurofilaments.

The cytoskeleton is also made up of certain motor proteins. These include:

Kinesin

These proteins move along the microtubules carrying the cellular components. They pull the organelles along the cell membrane.

Dyneins

These pull the cell organelles towards the nucleus.

Myosin

These interact with actin protein and are responsible for muscle contractions. They also perform cytokinesis, exocytosis, and endocytosis.

Cytoskeleton Functions

The important cytoskeleton functions are mentioned below:

1. It provides shape and support to the cell.
2. It helps in the formation of vacuoles.
3. It holds different cell organelles in place.
4. It assists in cell signalling.

5. It supports intracellular movements like the migration of cell organelles, transportation of vesicles in and out of the cell, etc.

Mitochondria

Mitochondria Definition

"Mitochondria are membrane-bound organelles present in the cytoplasm of all eukaryotic cells, that produces adinosine triphosphate (ATP), the main energy molecule used by the cell."

Popularly known as the "Powerhouse of the cell," mitochondria (singular: mitochondrion) are a double membrane-bound organelle found in most eukaryotic organisms. They are found inside the cytoplasm and essentially function as the cell's "digestive system."

They play a major role in breaking down nutrients and generating energy-rich molecules for the cell. Many of the biochemical reactions involved in cellular respiration take place within the mitochondria. The term 'mitochondrion' is derived from the Greek words "*mitos*" and "*chondrion*" which means "thread" and "granules-like", respectively. It was first described by a German pathologist named Richard Altmann in the year 1890.

The diagram of mitochondria below illustrates several structural features of mitochondria.

Structure of Mitochondria

- The mitochondrion is a double-membraned, rod-shaped structure found in both plant and animal cell.
- Its size ranges from 0.5 to 1.0 micrometre in diameter.
- The structure comprises an outer membrane, an inner membrane, and a gel-like material called the matrix.
- The outer membrane and the inner membrane are made of proteins and phospholipid layers separated by the intermembrane space.

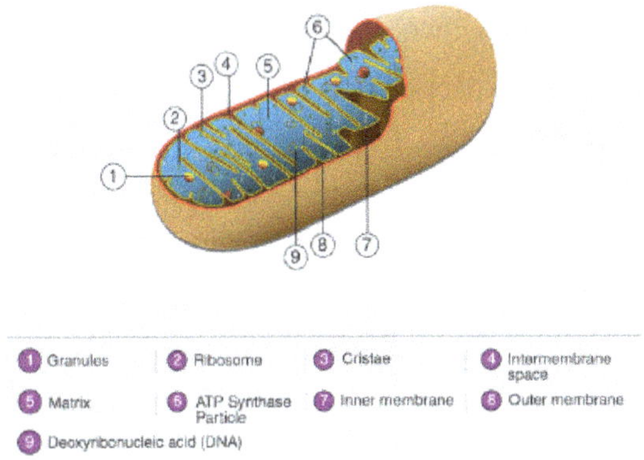

Fig. 2 Mitochondria diagram explaining the structure of mitochondria

- The outer membrane covers the surface of the mitochondrion and has a large number of special proteins known as porins.
- It is freely permeable to ions, nutrient molecules, energy molecules like the ADP and ATP molecules.

Cristae

The inner membrane of mitochondria is rather complex in structure. It has many folds that form a layered structure called cristae, and this helps in increasing the surface area inside the organelle. The cristae and the proteins of the inner membrane aids in the production of ATP molecules. The inner membrane is strictly permeable only to oxygen and to ATP molecules. A number of chemical reactions take place within the inner membrane of mitochondria.

Mitochondrial Matrix

The mitochondrial matrix is a viscous fluid that contains a mixture of enzymes and proteins. It also comprises ribosomes, inorganic ions, mitochondrial DNA, nucleotide cofactors, and organic molecules. The enzymes present in the matrix play an important role in the synthesis of ATP molecules.

Functions of Mitochondria

The most important function of mitochondria is to produce energy through the process of oxidative phosphorylation. It is also involved in the following process:

1. Regulates the metabolic activity of the cell
2. Promotes the growth of new cells and cell multiplication
3. Helps in detoxifying ammonia in the liver cells
4. Plays an important role in apoptosis or programmed cell death
5. Responsible for building certain parts of the blood and various hormones like testosterone and oestrogen
6. Helps in maintaining an adequate concentration of calcium ions within the compartments of the cell
7. It is also involved in various cellular activities like cellular differentiation, cell signalling, cell senescence, controlling the cell cycle and also in cell growth.

Disorders Associated With Mitochondria

Any irregularity in the way mitochondria functions can directly affect human health, but often, it is difficult to identify because symptoms differ from person to person. Disorders of the mitochondria can be quite severe; in some cases, it can even cause an organ to fail.

Mitochondrial diseases: Alpers Disease, Barth Syndrome, Kearns-Sayre syndrome (KSS)

Ribosomes

Ribosomes Definition

"Ribosomes are most important cell organelles composed of RNA and protein that converts genetic code into chains of amino acids."

A ribosome is a complex molecular machine found inside the living cells that produce proteins from amino acids during a process called protein synthesis or translation. The process of protein synthesis is a primary function, which is performed by all living cells.

Ribosomes are specialized cell organelles and are found in both prokaryotic and eukaryotic cells. Every living cell requires ribosomes for the production of proteins.

This cell organelle also functions by binding to a messenger ribonucleic acid (mRNA) and decoding the information carried by the nucleotide sequence of the mRNA. They transfer RNAs (tRNAs) comprising amino acids, and enter into the ribosome at the acceptor site. Once it gets bind up, it adds amino acid to the growing protein chain on tRNA.

The smaller subunit, where the mRNA binds and is decoded, and in the larger subunit, the amino acids get added. Both of the subunits contain both protein and ribonucleic acid components.

The two subunits are joined to each other by interactions between the rRNAs in one subunit and proteins in the other subunit.

Ribosomes are located inside the cytosol found in the plant cell and animal cell.

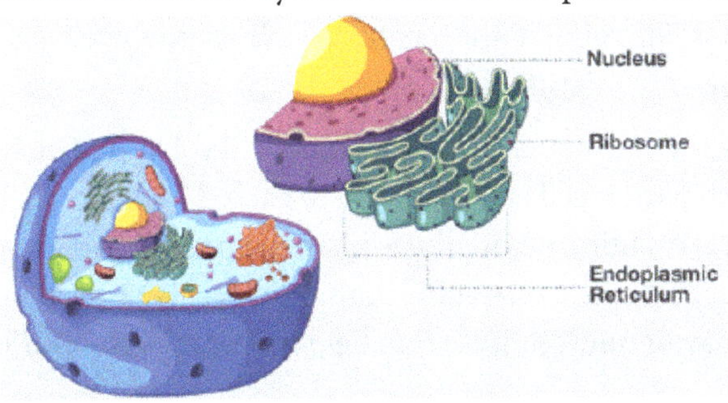

Fig. 3 Ribosomes Structure. A ribosome is a complex of RNA and protein and is, therefore, known as a ribonucleoprotein. It is composed of two subunits – smaller and larger.

The ribosome structure includes the following:

- It is located in two areas of cytoplasm.
- Scattered in the cytoplasm.
- Prokaryotes have 70S ribosomes while eukaryotes have 80S ribosomes.
- Around 62% of ribosomes are comprised of RNA, while the rest is proteins.

- The structure of free and bound ribosomes is similar and is associated with protein synthesis.

Ribosomes Function

The important ribosome function includes:

- It assembles amino acid to form proteins that are essential to carry out cellular functions.
- The DNA produces mRNA by the process of DNA transcription.
- The mRNA is synthesized in the nucleus and transported to the cytoplasm for the process of protein synthesis.
- The ribosomal subunits in the cytoplasm are bound around mRNA polymers. The tRNA then synthesizes proteins.
- The proteins synthesized in the cytoplasm are utilized in the cytoplasm itself, the proteins synthesized by bound ribosomes are transported outside the cell.

Centrosome

Centrosome Definition

"Centrosome is a mic Centrosome Structure rotubule-organizing centre in animal cells."

The centrosome is made up of two perpendicular centrioles, a daughter centriole, and a mother centriole, linked together by interconnecting fibres. It consists of a complex of proteins that helps in the formation of additional microtubules. An amorphous pericentriolar matrix surrounds the centrioles. It is involved in the nucleation and anchoring of cytoplasmic microtubules.

Centrosome in the animal cells is very much like DNA. During cell division, one centrosome from the parent cell is transferred to each daughter cell. In proliferating cells, the centrosome starts dividing before the S-phase begins. The newly formed centrosomes participate in organizing the mitotic spindles. During Interphase, the centrosome organizes an astral ray of microtubules that help in intracellular trafficking, cell adhesion, cell polarity, etc.

In post-mitotic cells, the centrosome consists of a mature centriole and an immature centriole, known as the mother centriole and daughter centriole respectively.

The centrosome cycle consists of four phases:

1. G1 phase where the duplication of centrosome takes place.
2. G2 phase where the centrosome maturation takes place.
3. The mitotic phase where the centrosome separation takes place.
4. A late mitotic phase where the chromosome disorientation takes place.

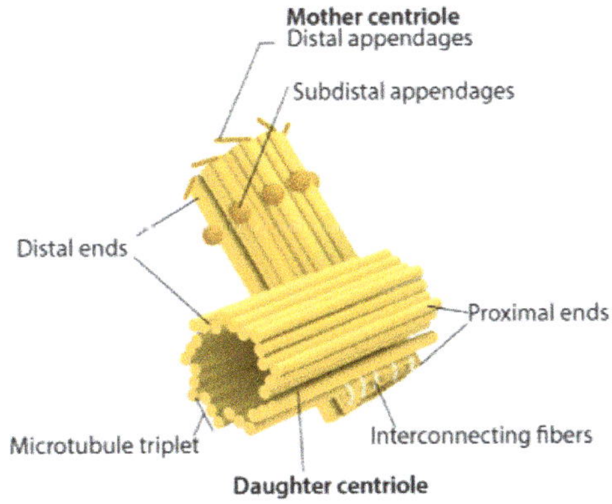

Fig. 4 Structure of Centrosome

Centrosome Function

The major functions of centrosome are listed below:

- The centrosomes help in cell division.
- They maintain the chromosome number during cell division.
- They also stimulate the changes in the shape of the cell membrane by phagocytosis.
- In mitosis, it helps in organizing the microtubules ensuring that the centrosomes are distributed to each daughter cell.
- They regulate the movement of microtubules and cytoskeletal structures, thereby, facilitating changes in the shapes of the membranes of the animal cell.

Centrosome in Animal Cells

In most animal cells, centrosomes are not required in the cell division process even though they add to the effectiveness of the mitotic spindle arrangement. In humans, dysfunctioning of centrosomes can stimulate cancer as a result of an increase in the levels of instability in chromosomes or due to the metastatic capability of cancer cells. However, the study on this lacks evidence.

Centrosome in Plant Cells

Plants and fungi do not possess centrosomes hence make use of MTOC structures to coordinate their microtubules. Plant cells do not possess spindle pole bodies or centrioles except in flagellate male gametes which are completely present in a few flowering plants(conifers). The primary function of the MTOC for spindle organization and microtubule nucleation appears to be taken up by the nuclear envelope while the mitosis of the plant cell.

A rare pathway has been developed by higher plants to regulate the dynamics and assembly of the cytoskeleton. The microtubules are nucleated at the organizing and nucleation centres in many other eukaryotes which are committed to establishing polarity. Though animal and plant cells share the main cytoskeleton elements that imply controlled working, plants do not exhibit centrosome resembling organelles but yet are capable of building spindles thus have developed cytoskeletal arrays such as the preprophase band, the cortical arrays and the phragmoplast that participate in fundamental growth processes.

Certain elements such as the gamma-tubulin etc have a major role to play in the microtubule nucleation taking place at the surface of the nucleus which is referred to as the major operative plant microtubule-organizing centre.

Plastids

Plastids are double-membrane organelles which are found in the cells of plants and algae. Plastids are responsible for manufacturing and storing of food. These often contain pigments that are used in photosynthesis and different types of pigments that can change the colour of the cell.

Types of Plastids

There are different types of plastids with their specialized functions. Among them, a few are mainly classified based on the presence or absence of the Biological pigments and their stages of development.

- Chloroplasts
- Chromoplasts
- Gerontoplasts
- Leucoplasts

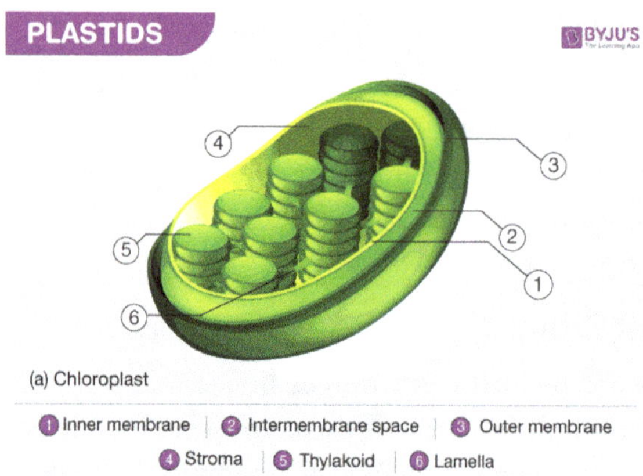

Fig. 5 Plastids

Chloroplasts

Chloroplasts are biconvex shaped, semi-porous, double membraned, cell organelle found within the mesophyll of the plant cell. They are the sites for synthesizing food by the process of photosynthesis.

Chromoplasts

Chromoplasts is the name given to an area for all the pigments to be kept and synthesized in the plant. These can be usually found in flowering plants, ageing leaves and fruits. Chloroplasts convert into chromoplasts. Chromoplasts have carotenoid pigments that allow different colours that you see in leaves and fruits. The main reason for its different colour is for attracting pollinators.

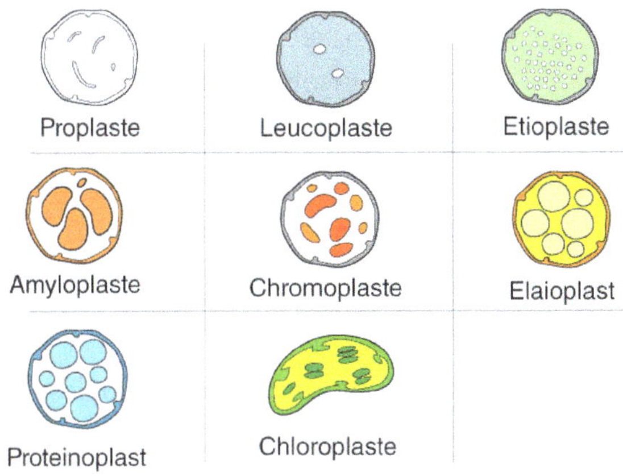

Fig. 6 Types of Plastids

Gerontoplasts

These are basically chloroplasts that go with the ageing process. Geronoplasts refer to the chloroplasts of the leaves that help to convert into different other organelles when the leaf is no longer using photosynthesis usually in an autumn month.

Leucoplasts

These are the non-pigmented organelles which are colourless. Leucoplasts are usually found in most of the non-photosynthetic parts of the plant like roots. They act as a storage sheds for starches, lipids, and proteins depending on the need of the plants. They are mostly used for converting amino acids and fatty acids.

Leucoplasts are of three types:

- Amyloplasts – Amyloplasts are greatest among all three and they store and synthesize starch.
- Proteinoplasts – Proteinoplasts help in storing the proteins that a plant needs and can be typically found in seeds.
- Elaioplasts -Elaioplast helps in storing fats and oils that are needed by the plant.

Inheritance of Plastids

There are many plants which are inherited from the plastids from just a single parent. Angiosperms inherit plastids from the female gamete while there are many gymnosperms that inherit plastids from the male pollen. Algae inherit plastids from one parent only. The inheritance of the plastids-DNA seems to be 100% uniparental. In hybridisation, the inheritance of plastid seems to be more erratic.

Vacuoles

"Vacuoles are membrane-bound cell organelles present in the cytoplasm and filled with a watery fluid containing various substances."

The term "vacuole" means "empty space". They help in the storage and disposal of various substances. They can store food or other nutrients required by a cell to survive. They also store waste products and prevent the entire cell from contamination.

The vacuoles in plant cells are larger than those in the animal cells. The plant vacuoles occupy more than 80% of the volume of the cell. The vacuoles may be one or more in number.

Structure of Vacuole

A vacuole is a membrane bound structure found in the cytoplasmic matrix of a cell. The membrane surrounding the vacuole is known as tonoplast. The components of the vacuole, known as the cell sap, differ from that of the surrounding cytoplasm. The membranes are composed of phospholipids. The membranes are embedded with proteins that help in transporting molecules across the membrane. Different combinations of these proteins help the vacuoles to hold different materials.

Functions of Vacuole

The important functions of vacuole include:

Storage

A vacuole stores salts, minerals, pigments and proteins within the cell. The solution that fills a vacuole is known as the cell sap. The vacuole is also filled with protons from the cytosol that helps in maintaining an acidic environment within the cell. A large number of lipids are also stored within the vacuoles.

Turgor Pressure

The vacuoles are completely filled with water and exert force on the cell wall. This is known as turgor pressure. It provides shape to the cell and helps it to withstand extreme conditions.

Endocytosis and Exocytosis

The substances are taken in by a vacuole through endocytosis and excreted through exocytosis. These substances are stored in the cells, separated from the cytosol. Lysosomes are vesicles that intake food and digest it. This is endocytosis and it varies in different cells.

Prokaryotic Cells

"Prokaryotic cells are the cells that do not have a true nucleus and membrane-bound organelles."

Prokaryotic cells are single-celled microorganisms known to be the earliest on earth. Prokaryotes include Bacteria and Archaea. The photosynthetic prokaryotes include cyanobacteria that perform photosynthesis.

A prokaryotic cell consists of a single membrane and therefore, all the reactions occur within the cytoplasm. They can be free-living or parasites.

Characteristics of Prokaryotic Cell

Prokaryotic cells have different characteristic features. The characteristics of the prokaryotic cells are mentioned below.

1. They lack a nuclear membrane.
2. Mitochondria, Golgi bodies, chloroplast, and lysosomes are absent.
3. The genetic material is present on a single chromosome.
4. The histone proteins, the important constituents of eukaryotic chromosomes, are lacking in them.
5. The cell wall is made up of carbohydrates and amino acids.
6. The plasma membrane acts as the mitochondrial membrane carrying respiratory enzymes.
7. They divide asexually by binary fission. The sexual mode of reproduction involves conjugation.

Prokaryotic Cell Structure

A prokaryotic cell does not have a nuclear membrane. However, the genetic material is present in a region in the cytoplasm known as the nucleoid. They may be spherical, rod-shaped, or spiral. A prokaryotic cell structure is as follows:

1. **Capsule** – It is an outer protective covering found in the bacterial cells, in addition to the cell wall. It helps in moisture retention, protects the cell when engulfed, and helps in the attachment of cells to nutrients and surfaces.
2. **Cell Wall** – It is the outermost layer of the cell which gives shape to the cell.
3. **Cytoplasm** – The cytoplasm is mainly composed of enzymes, salts, cell organelles and is a gel-like component.
4. **Cell Membrane** – This layer surrounds the cytoplasm and regulates the entry and exit of substances in the cells.
5. **Pili** – These are hair-like outgrowths that attach to the surface of other bacterial cells.
6. **Flagella** – These are long structures in the form of a whip, that help in the locomotion of a cell.
7. **Ribosomes** – These are involved in protein synthesis.

8. **Plasmids** – Plasmids are non-chromosomal DNA structures. These are not involved in reproduction.
9. **Nucleoid Region** – It is the region in the cytoplasm where the genetic material is present.

A prokaryotic cell lacks certain organelles like mitochondria, endoplasmic reticulum, and Golgi bodies.

The prokaryotic cell diagram given below represents a bacterial cell. It depicts the absence of a true nucleus and the presence of a flagellum that differentiates it from a eukaryotic cell.

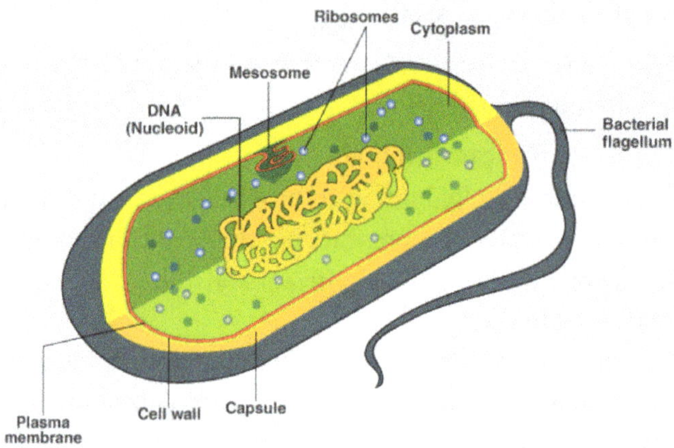

Fig. 7. Prokaryotic Cell Diagram illustrates the absence of a true nucleus

Components of Prokaryotic Cells

The prokaryotic cells have four main components:

Plasma Membrane - It is an outer protective covering of phospholipid molecules which separates the cell from the surrounding environment.

Cytoplasm - It is a jelly-like substance present inside the cell. All the cell organelles are suspended in it.

DNA - It is the genetic material of the cell. All the prokaryotes possess a circular DNA. It directs what proteins the cell creates. It also regulates the actions of the cell.

Ribosomes - Protein synthesis occurs here

Some prokaryotic cells possess cilia and flagella which helps in locomotion.

Reproduction in Prokaryotes

A prokaryote reproduces in two ways:

- Asexually by binary fission
- Sexually by conjugation

Binary Fission

1. The DNA of an organism replicates and the new copies attach to the cell membrane.
2. The cell wall starts increasing in size and starts moving inwards.
3. A cell wall is then formed between each DNA, dividing the cell into two daughter cells.

Recombination

In this process, genes from one bacteria are transferred to the genome of other bacteria. It takes place in three ways-conjugation, transformation, transduction.

- Conjugation is the process in which genes are transferred between two bacteria through a protein tube structure called a pilus.
- Transformation is the mode of sexual reproduction in which the DNA from the surroundings is taken by the bacterial cell and incorporated in its DNA.
- Transduction is the process in which the genetic material is transferred into the bacterial cell with the help of viruses. Bacteriophages are the virus that initiates the process.

Examples of Prokaryotic Cells

The examples of the prokaryotic cells are mentioned below:

Bacterial Cells

These are unicellular organisms found everywhere on earth from soil to the human body.

They have different shapes and structures.

The cell wall is composed of peptidoglycan that provides structure to the cell wall.

Bacteria have some unique structures such as pili, flagella and capsule.

They also possess extrachromosomal DNA known as plasmids.

They have the ability to form tough, dormant structures known as endospores that helps them to survive under unfavourable conditions. The endospores become active when the conditions are favourable again.

Archaeal Cells

Archaebacteria are unicellular organisms similar to bacteria in shape and size.

They are found in extreme environments such as hot springs and other places such as soil, marshes, and even inside humans.

They have a cell wall and flagella. The cell wall of archaea does not contain peptidoglycan.

The membranes of the archaea have different lipids with a completely different stereochemistry.

Just like bacteria, archaea have one circular chromosome. They also possess plasmids.

For more information on Prokaryotic Cells, its definition, structure, characteristics and examples, keep visiting BYJU'S Biology website or download BYJU'S app for further reference.

Endoplasmic Reticulum

"Endoplasmic Reticulum is a complex network of tubular membranes exclusively present in the cytoplasm of the eukaryotic cell."

The endoplasmic reticulum transpires in two forms: a type with a ribosome-studded surface and another with a smooth surface. The latter is called the smooth endoplasmic reticulum, and the former is called the rough endoplasmic reticulum. These membranes form continuous folds, eventually joining the outer layer of the nuclear membrane. Except for sperm cells and red blood cells, the endoplasmic reticulum is observed in every other type of eukaryotic cell.

The below diagram shows the variants of the endoplasmic reticulum:

- Rough ER
- Smooth ER

Rough endoplasmic reticulum has ribosomes embedded within its structure, giving a "rough" appearance. The smooth endoplasmic reticulum does not have these ribosomes, hence appearing "smooth."

Structure of Endoplasmic Reticulum

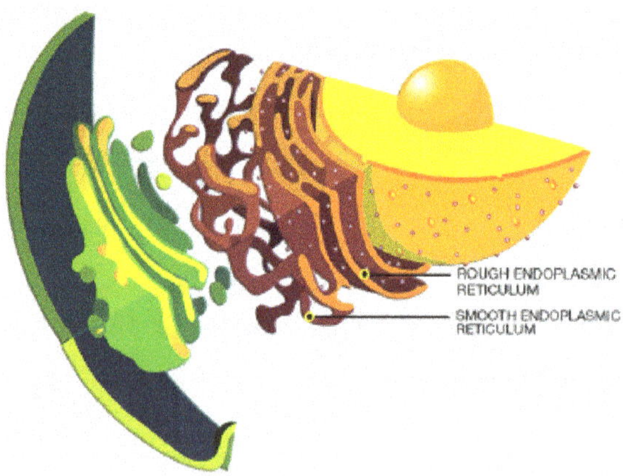

Fig 8. Endoplasmic Reticulum

The structure of the endoplasmic reticulum is shaped like a sac. Since ER is of two types, each has its own distinguishing features:

Rough Endoplasmic Reticulum Structure

- The rough endoplasmic reticulum is named so because of its appearance.
- It is a series of connected flattened sacs having several ribosomes on its outer surface, hence the name.

- It synthesizes and secretes proteins in the liver, hormones and other substances in the glands.
- Rough ER is prominent in cells where protein synthesis happens (such as hepatocytes)

Smooth Endoplasmic Reticulum Structure

- The smooth endoplasmic reticulum, on the other hand, does not have ribosomes.
- The smooth endoplasmic reticulum has a tubular form.
- It participates in the production of phospholipids, the chief lipids in cell membranes and are essential in the process of metabolism.
- Smooth ER transports the products of the rough ER to other cellular organelles, especially the Golgi apparatus.

Functions of Endoplasmic Reticulum

As stated above, the endoplasmic reticulum is categorised into two types, and both these types of ER perform specific functions:

Smooth Endoplasmic Reticulum Functions:

- Smooth ER is responsible for the synthesis of essential lipids such as phospholipids and cholesterol.
- Smooth ER is also responsible for the production and secretion of steroid hormones.
- It is also responsible for the metabolism of carbohydrates.
- The smooth ER store and releases calcium ions. These are quite important for the nervous system and muscular systems.

Rough Endoplasmic Reticulum Functions:

- The majority of the functions of rough ER is associated with protein synthesis.
- The rough endoplasmic reticulum also plays a vital role in protein folding.
- Also ensures quality control (regarding correct protein folding).
- The second most important function after protein synthesis and protein folding is protein sorting.

Golgi Apparatus

The Golgi apparatus has multiple names such as Golgi complex or Golgi body. The name is given on the name of the scientist, who discovered the organelle, i.e. Camillo Golgi. It is found in all the eukaryotic cells, plants as well as animals. They are membrane-bound organelle present in the cytosol of the cell. Let us explore more about Golgi complex.

Golgi Apparatus

The Golgi body comprises 5 to 8 cup-shaped, series of compartments known as cisternae. Cisternae is a flattened, disk-shaped, stacked pouches that make up the Golgi apparatus. A Golgi stack mostly contains 4 to 8 cisternae. However, ~60 cisternae are found in some protists. A mammalian cell contains ~40 to 100 stacks of cisternae.

Animal cells generally contain around 10 to 20 Golgi stacks per cell, which are connected by tubular connections. Golgi complex is mostly found near the nucleus.

Creation, or evolution, whichever one, you hold a belief in has worked in wondrous ways to evolve or design the various living beings in this world in the most optimum ways. For example, take the Golgi complex, it has been designed in such a way, to ensure a sufficient number of Golgi bodies are present in the cell as per the requirement.

Golgi Bodies Functions

Its main function is the packaging and secretion of proteins. It receives proteins from Endoplasmic Reticulum. It packages it into membrane-bound vesicles, which are then transported to various destinations, such as lysosomes, plasma membrane or secretion. They also take part in the transport of lipids and the formation of lysosomes.

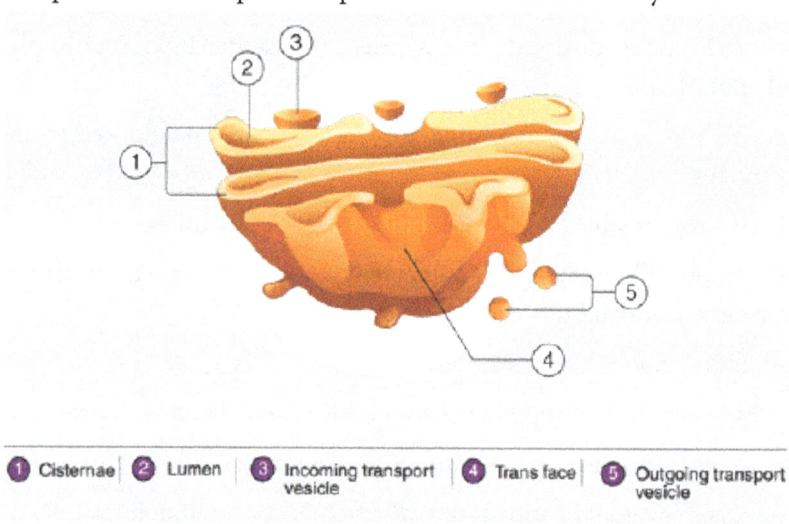

Fig. 9 The Golgi apparatus

Post-translational modification and enzymatic processing occur near the membrane surface in Golgi bodies, e.g. phosphorylation, glycosylation, etc.

Golgi apparatus is the site for the synthesis of various glycolipids, sphingomyelin, etc.

In the plant cells, complex polysaccharides of the cell wall are synthesised in the Golgi apparatus.

Nucleus - Structure and Function

The nucleus is a pivotal organelle responsible for regulating almost all forms of cellular activities. Mostly, every type of cell that exists is categorized on the basis of the absence or presence of the nucleus within its cell (categorized either as a prokaryotic or eukaryotic cell.)

The most integral component of the cell is the nucleus (plural: nuclei). It is derived from a Latin word which means "*kernel of a nut*".

Nucleus Definition

A nucleus is defined as a double-membraned eukaryotic cell organelle that contains the genetic material.

As stated above, the nucleus is found only in eukaryotes and is the defining characteristic feature of eukaryotic cells. However, some cells, such as RBCs do not possess a nucleus, though they originate from a eukaryotic organisms.

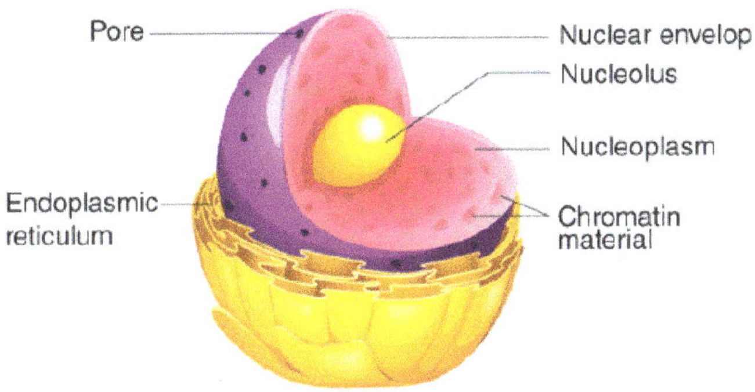

Fig. 10 A nucleus diagram highlighting the various components.

Structure of Nucleus

- Typically, it is the most evident organelle in the cell.
- The nucleus is completely bound by membranes.
- It is engirdled by a structure referred to as the nuclear envelope.
- The membrane distinguishes the cytoplasm from the contents of the nucleus
- The cell's chromosomes are also confined within it.
- DNA is present in the Chromosomes, and they provide the genetic information required for the creation of different cell components in addition to the reproduction of life.

Nucleus Functions

Following are the important nucleus function:

- It contains the cell's hereditary information and controls the cell's growth and reproduction.

- The nucleus has been clearly explained as a membrane-bound structure that comprises the genetic material of a cell.
- It is not just a storage compartment for DNA, but also happens to be the home of some important cellular processes.
- First and foremost, it is possible to duplicate one's DNA in the nucleus. This process has been named DNA Replication and produces an identical copy of the DNA.
- Producing two identical copies of the body or host is the first step in cell division, where every new cell will get its own set of instructions.
- Secondly, the nucleus is the site of transcription. Transcription creates different types of RNA from DNA. Transcription would be a lot like creating copies of individual pages of the human body's instructions which may be moved out and read by the rest of the cell.
- The central rule of biology states that DNA is copied into RNA, and then proteins.

Microbodies

A microbody is a cell organelle present in both plant and animal cells. Glyoxysomes, peroxisomes are included in the microbodies family. In Vertebrates, microbodies are prevalent in the kidney and liver cells.

Structure

Microbodies are present in the cytoplasm of a cell so they are known as cytosomes. They are very small in size ~0.2-1.5 μm and can be seen under the electron microscope.

Some of the common characteristics of microbodies are:

- They are vesicular and mostly spherical in shape
- They are enclosed in a single membrane of a phospholipid bilayer
- The intracellular matrix contains proteins and enzymes
- They do not contain separate DNA

Functions

Different types of microbodies perform various specific functions.

- Microbodies take part in various biochemical reactions in the cell
- The enzymes present in microbodies facilitate various essential reactions, e.g. breakdown of fats, amino acids, alcohol, etc.
- They are involved in the photorespiration in plants
- Detoxification of peroxides occurs in microbodies

Peroxisomes

Peroxisomes are a membrane-bound organelle present in the eukaryotic cells. They take part in various oxidative processes. They take part in lipid metabolism and catabolism of D-amino acids, polyamines and bile acids. The reactive oxygen species such as peroxides produced in the process is converted to water by various enzymes like peroxidase and catalase.

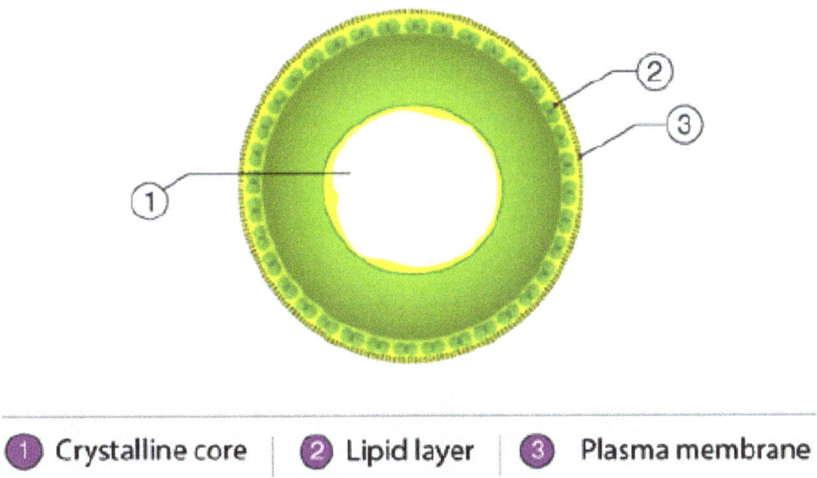

Fig. 11 Microbodies

In plants, photorespiration occurs in peroxisomes.

Glyoxysomes

Glyoxysomes are specialized peroxisomes. Their main function is to convert fatty acid to carbohydrate. They are present in plants and fungi. They are prevalent in the germinating seeds in their fat-storing tissues.

3

MENDELISM AND CHROMOSOMAL ABERRATION

Mendelian inheritance, also called Mendelism, the principles of heredity formulated by Austrian-born botanist, teacher, and Augustinian prelate Gregor Mendel in 1865. These principles compose what is known as the system of particulate inheritance by units, or genes. The later discovery of chromosomes as the carriers of genetic units supported Mendel's two basic laws, known as the law of segregation and the law of independent assortment.

In modern terms, the first of Mendel's laws states that genes are transferred as separate and distinct units from one generation to the next. The two members (alleles) of a gene pair, one on each of paired chromosomes, separate during the formation of sex cells by a parent organism. One-half of the sex cells will have one form of the gene, one-half the other form; the offspring that result from these sex cells will reflect those proportions.

The inheritance of traits from parents to the offsprings in a dominant-recessive manner is called mendelism.

Law of Independent Assortment: When plants with two or more than two characters are crossed then expression of one character does not affect the expression of another character. Genes from a pair of alleles segregate and are free to combine with any other gene from any other pair of alleles. This is called law of independent assortment.

Example: The dihybrid cross between a plant with round and green seeds (RRyy) and another plant with wrinkled and yellow seeds (rrYY). Here roundness or smoothness of seeds and yellow colour are dominant traits, while wrinkles on seeds and green colour are recessive traits.

When these plants were crossed then all plants in F_1 generation produced round and yellow seeds. When plants of F_1 generation were allowed to reproduce, plants of F_2 generation showed following traits in respective ratio:

Round yellow : Round green : Wrinkled yellow : Wrinkled green

9 : 3 : 3 :1

The observation that roundness combines with either of the colours and roughness combines with either of the colours proves the law of independent assortment.

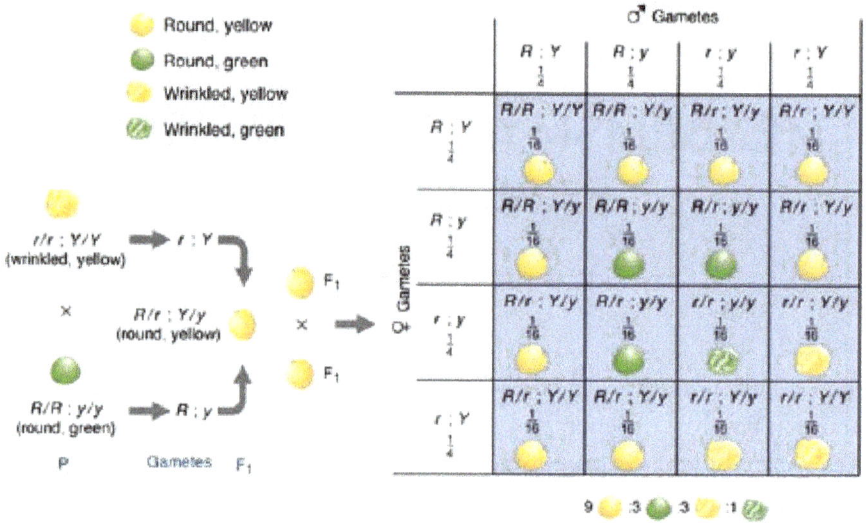

Fig. 1

Mendelism is used to refer to the theoretical principles of heredity of the single-gene trait, which is derived from the principle put forward by Gregor Mendel, known as Mendel's laws.

Mendel's laws came to be known as Laws of inheritance, which are as follows:

- Law of Dominance
- Law of Segregation
- Law of Independent Assortment

MENDEL'S LAWS OF INHERITANCE

Inheritance can be defined as the process of how a child receives genetic information from the parent. The whole process of heredity is dependent upon inheritance and it is the reason that the offsprings are similar to the parents. This simply means that due to inheritance, the members of the same family possess similar characteristics.

It was only during the mid 19th century that people started to understand inheritance in a proper way. This understanding of inheritance was made possible by a scientist named Gregor Mendel, who formulated certain laws to understand inheritance known as Mendel's laws of inheritance.

Mendel's Laws of Inheritance

Between 1856-1863, Mendel conducted the hybridization experiments on the garden peas. During that period, he chose some distinct characteristics of the peas and conducted some cross-pollination/ artificial pollination on the pea lines that showed stable trait inheritance and underwent continuous self-pollination. Such pea lines are called true-breeding pea lines.

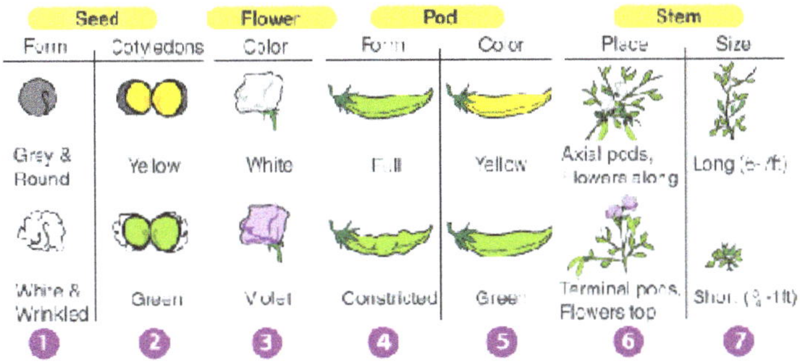

Fig. 2. Mendel's Laws of Inheritance

Why was Pea Plant Selected for Mendel's Experiments?

He selected a pea plant for his experiments:

1. The pea plant can be easily grown and maintained.
2. They are naturally self-pollinating but can also be cross-pollinated.
3. It is an annual plant, therefore, many generations can be studied within a short period of time.
4. It has several contrasting characters.

Mendel conducted 2 main experiments to determine the laws of inheritance. These experiments were:

1. Monohybrid Cross Experiment
2. Dihybrid Cross Experiment

While experimenting, Mendel found that certain factors were always being transferred down to the offspring in a stable way. Those factors are now called genes i.e. genes can be called the units of inheritance.

MENDEL'S EXPERIMENTS

Mendel experimented on a pea plant and considered 7 main contrasting traits in the plants. Then, he conducted both the experiments to determine the aforementioned inheritance laws. A brief explanation of the two experiments is given below.

Monohybrid Cross

In this experiment, Mendel took two pea plants of opposite traits (one short and one tall) and crossed them. He found the first generation offsprings were tall and called it F1 progeny. Then he crossed F1 progeny and obtained both tall and short plants in the ratio 3:1. To know more about this experiment, visit Monohybrid Cross – Inheritance Of One Gene.

Mendel even conducted this experiment with other contrasting traits like green peas vs yellow peas, round vs wrinkled, etc. In all the cases, he found that the results were similar. From this, he formulated the laws of Segregation and Dominance.

Dihybrid Cross

In a dihybrid cross experiment, Mendel considered two traits, each having two alleles. He crossed wrinkled-green seed and round-yellow seeds and observed that all the first generation progeny (F1 progeny) were round-yellow. This meant that dominant traits were the round shape and yellow colour.

He then self-pollinated the F1 progeny and obtained 4 different traits wrinkled-yellow, round-yellow, wrinkled-green seeds and round-green in the ratio 9:3:3:1.

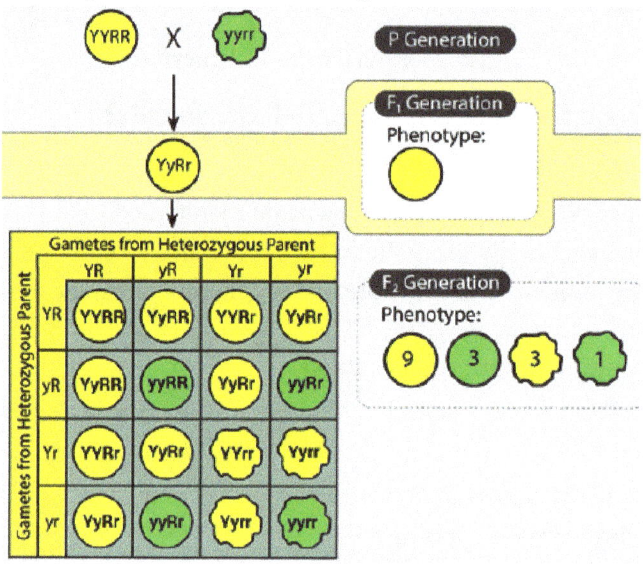

Fig. 4. Dihybrid Cross

After conducting for other traits, the results were found to be similar. From this experiment, Mendel formulated his second law of inheritance i.e law of Independent Assortment.

CONCLUSIONS FROM MENDEL'S EXPERIMENTS

- The genetic makeup of the plant is known as the genotype. On the contrary, the physical appearance of the plant is known as phenotype
- The genes are transferred from parents to the offsprings in pairs known as allele.
- During gametogenesis when the chromosomes are halved, there is a 50% chance of one of the two alleles to fuse with the other parent.

- When the alleles are the same, they are known as homozygous alleles and when the alleles are different they are known as heterozygous alleles.

MENDEL'S LAWS

The two experiments lead to the formulation of Mendel's laws known as laws of inheritance which are:

1. Law of Dominance
2. Law of Segregation
3. Law of Independent Assortment

Law of Dominance

This is also called Mendel's first law of inheritance. According to the law of dominance, hybrid offsprings will only inherit the dominant trait in the phenotype. The alleles that are suppressed are called as the recessive traits while the alleles that determine the trait are known as the dormant traits.

Law of Segregation

The law of segregation states that during the production of gametes, two copies of each hereditary factor segregate so that offspring acquire one factor from each parent. In other words, allele (alternative form of the gene) pairs segregate during the formation of gamete and re-unite randomly during fertilization. This is also known as Mendel's third law of inheritance.

Law of Independent Assortment

Also known as Mendel's second law of inheritance, the law of independent assortment states that a pair of trait segregates independently of another pair during gamete formation. As the individual heredity factors assort independently, different traits get equal opportunity to occur together.

Key Points on Mendel's Laws

- The law of inheritance was proposed by Gregor Mendel after conducting experiments on pea plants for seven years.
- The Mendel's laws of inheritance include law of dominance, law of segregation and law of independent assortment.
- The law of segregation states that every individual possesses two alleles and only one allele is passed on to the offspring.
- The law of independent assortment states that the inheritance of one pair of genes is independent of inheritance of another pair.

Mendel's 3 Principles?

Mendel's three principles are –

- **Law of Dominance** – When there are two different alleles, one is dominant and other recessive
- **Law of Segregation** – two alleles for each gene are placed in different gametes
- **Law of Independent Assortment** – the inheritance of one gene does not affect the inheritance of another gene

Mendel's first and second law

Mendel's first law is the law of dominance and Mendel's second law is the law of segregation.

Mendel's First Law

The first law of Mendel is the law of segregation that describes the separation of the two copies of each hereditary factor or gene during the formation of gametes. Each gene exists in two copies called the alleles within a diploid genome. Each allele comes from each parent. During the formation of gametes, the allele pair separates from each other in such a way that each gamete receives one allele from the pair. Therefore, the offspring acquires one copy from each parent. During the fusion of gametes, it acquires two alleles from each parental gamete.

Here, alleles can be either homozygous or heterozygous. One allele in the heterozygous pair is dominant while the other is recessive. The expression of the dominant allele to produce the phenotype is called complete dominance. Shown in Fig. 4 is the punnett square describing Mendel's first law by a monohybrid cross.

The phenotypic ratio between the offspring in Mendel's first law is 3:1.

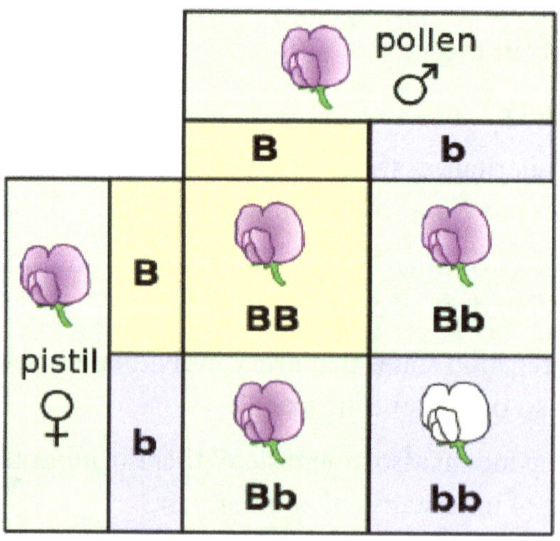

Fig. 4: Mendel's First Law

Mendel's Second Law

The second law of Mendel is the law of independent assortment. This law describes how alleles of different genes independently segregate from each other during the formation of gametes. However, it is applicable only when two or more factors are inherited together. According to Mendel's second law, the trans-mission of different alleles of different genes into the gametes is not affected by each other. Mendel describes the second law with a dihybrid cross. The phenotypic ratio of a dihybrid cross is 9:3:3:1. Figure 5 below shows the behavior of the characters; short tail (S), long tail (s) brown coat (B), and white coat (b) in a dihybrid cross.

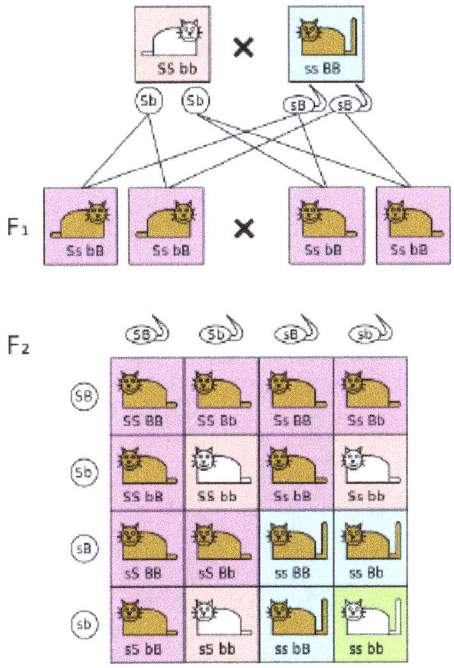

Fig. 5: Mendel's Second Law

Independent assortment occurs during prophase 1 of meiosis 1 in the formation of gametes. The random orientation of the bivalent chromosomes on the equatorial plate of the cell during the metaphase 1 is the physical basis of the independent assortment. Genetic linkage violates Mendel's second law.

Mendel's third law

Mendel's third law is the law of independent assortment

- The genotype of an individual is made up of the many alleles it possesses.
- An individual's physical appearance, or phenotype, is determined by its alleles as well as by its environment.
- The presence of an allele does not mean that the trait will be expressed in the individual that possesses it.

- If the two alleles of an inherited pair differ (the heterozygous condition), then one determines the organism's appearance and is called the dominant allele; the other has no noticeable effect on the organism's appearance and is called the recessive allele.
- Thus, the dominant allele will hide the phenotypic effects of the recessive allele.
- This is known as the Law of Dominance but it is not a transmission law: it concerns the expression of the genotype.

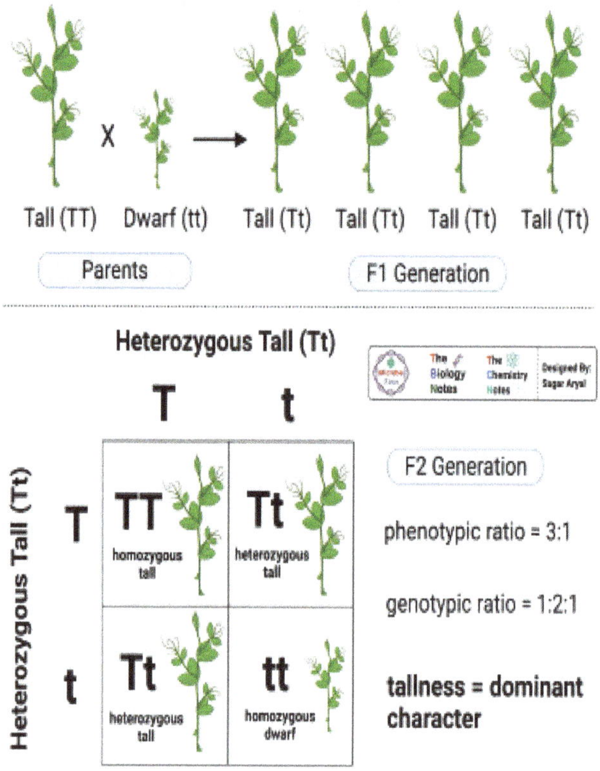

Fig. 6 Mendal's Law of Dominance

- The upper case letters are used to represent dominant alleles whereas the lowercase letters are used to represent recessive alleles.

CHROMOSOMAL ABERRATION

Chromosomal aberrations, or abnormalities, are changes to the structure or number of chromosomes, which are strands of condensed genetic material. Humans typically have 23 pairs of chromosomes, of which 22 pairs are autosomal, numbered 1 through 22. The last pair of chromosomes are sex chromosomes, which determine an individual's sex assignment. At birth, most people with XY sex chromosomes are assigned male, and most individuals with XX are assigned female. In general, each parent contributes one set of chromosomes to their offspring, which collectively make up the 23 pairs of chromosomes.

A change to any of the chromosomes, in number or structure, creates a chromosomal aberration and may cause medical disorders.

Changes in chromosome structure

Two important principles dictate the properties of a large proportion of structural chromosomal changes. The first principle is that any deviation from the normal ratio of genetic material in the genome results in genetic imbalance and abnormal function. In the normal nuclei of both diploid and haploid cells, the ratio of the individual chromosomes to one another is 1:1. Any deviation from this ratio by addition or subtraction of either whole chromosomes or parts of chromosomes results in genomic imbalance. The second principle is that homologous chromosomes go to great lengths to pair at meiosis. The tightly paired homologous regions are joined by a ladderlike longitudinal structure called the synaptonemal complex. Homologous regions seem to be able to find each other and form a synaptonemal complex whether or not they are part of normal chromosomes. Therefore, when structural changes occur, not only are the resulting pairing formations highly characteristic of that type of structural change but they also dictate the packaging of normal and abnormal chromosomes into the gametes and subsequently into the progeny.

Deletions

The simplest, but perhaps most damaging, structural change is a deletion—the complete loss of a part of one chromosome. In a haploid cell this is lethal, because part of the essential genome is lost. However, even in diploid cells deletions are generally lethal or have other serious consequences. In a diploid a heterozygous deletion results in a cell that has one normal chromosome set and another set that contains a truncated chromosome. Such cells show genomic imbalance, which increases in severity with the size of the deletion. Another potential source of damage is that any recessive, deleterious, or lethal alleles that are in the normal counterpart of the deleted region will be expressed in the phenotype. In humans, cri-du-chat syndrome is caused by a heterozygous deletion at the tip of the short arm of chromosome 5. Infants are born with this condition as the result of a deletion arising in parental germinal tissues or even in sex cells. The manifestations of this deletion, in addition to the "cat cry" that gives the syndrome its name, include severe intellectual disability and an abnormally small head.

Duplications

A heterozygous duplication (an extra copy of some chromosome region) also results in a genomic imbalance with deleterious consequences. Small duplications within a gene can arise spontaneously. Larger duplications can be caused by crossovers following asymmetrical chromosome pairing or by meiotic irregularities resulting from other types of altered chromosome structures. If a duplication becomes homozygous, it can provide the organism with an opportunity to acquire new genetic functions through mutations within the duplicate copy.

Inversions

An inversion occurs when a chromosome breaks in two places and the region between the break rotates 180° before rejoining with the two end fragments. If the inverted segment contains the centromere (i.e., the point where the two chromatids are joined), the inversion is said to be pericentric; if not, it is called paracentric. Inversions do not result in a gain or loss of genetic material, and they have deleterious effects only if one of the chromosomal breaks occurs within an essential gene or if the function of a gene is altered by its relocation to a new chromosomal neighbourhood (called the position effect). However, individuals who are heterozygous for inversions produce aberrant meiotic products along with normal products. The only way uninverted and inverted segments can pair is by forming an inversion loop. If no crossovers occur in the loop, half of the gametes will be normal and the other half will contain an inverted chromosome. If a crossover does occur within the loop of a paracentric inversion, a chromosome bridge and an acentric chromosome (i.e., a chromosome without a centromere) will be formed, and this will give rise to abnormal meiotic products carrying deletions, which are inviable. In a pericentric inversion, a crossover within the loop does not result in a bridge or an acentric chromosome, but inviable products are produced carrying a duplication and a deletion.

Translocations

If a chromosome break occurs in each of two nonhomologous chromosomes and the two breaks rejoin in a new arrangement, the new segment is called a translocation. A cell bearing a heterozygous translocation has a full set of genes and will be viable unless one of the breaks causes damage within a gene or if there is a position effect on gene function. However, once again the pairing properties of the chromosomes at meiosis result in aberrant meiotic products. Specifically, half of the products are deleted for one of the chromosome regions that changed positions and half of the products are duplicated for the other. These duplications and deletions usually result in inviability, so translocation heterozygotes are generally semisterile ("half-sterile").

Changes in chromosome number

Two types of changes in chromosome numbers can be distinguished: a change in the number of whole chromosome sets (polyploidy) and a change in chromosomes within a set (aneuploidy).

Polyploids

An individual with additional chromosome sets is called a polyploid. Individuals with three sets of chromosomes (triploids, $3n$) or four sets of chromosomes (tetraploids, $4n$) are polyploid derivatives of the basic diploid ($2n$) constitution. Polyploids with odd numbers of sets (e.g., triploids) are sterile, because homologous chromosomes pair only two by two, and the extra chromosome moves randomly to a cell pole, resulting in highly unbalanced, nonfunctional meiotic products. It is for this reason that triploid watermelons are seedless. However, polyploids with even numbers of chromosome sets can be fertile if orderly two-by-two chromosome pairing occurs.

Though two organisms from closely related species frequently hybridize, the chromosomes of the fusing partners are different enough that the two sets do not pair at meiosis, resulting in sterile offspring. However, if by chance the number of chromosome sets in the hybrid accidentally duplicates, a pairing partner for each chromosome will be produced, and the hybrid will be fertile. These chromosomally doubled hybrids are called allotetraploids. Bread wheat, which is hexaploid ($6n$) due to several natural spontaneous hybridizations, is an example of an allotetraploid. Some polyploid plants are able to produce seeds through an asexual type of reproduction called apomixis; in such cases, all progeny are identical to the parent. Polyploidy does arise spontaneously in humans, but all polyploids either abort in utero or die shortly after birth.

Aneuploids

Some cells have an abnormal number of chromosomes that is not a whole multiple of the haploid number. This condition is called aneuploidy. Most aneuploids arise by nondisjunction, a failure of homologous chromosomes to separate at meiosis. When a gamete of this type is fertilized by a normal gamete, the zygotes formed will have an unequal distribution of chromosomes. Such genomic imbalance results in severe abnormalities or death. Only aneuploids involving small chromosomes tend to survive and even then only with an aberrant phenotype.

The most common form of aneuploidy in humans results in Down syndrome, a suite of specific disorders in individuals possessing an extra chromosome 21 (trisomy 21). The symptoms of Down syndrome include intellectual disability, severe disorders of internal organs such as the heart and kidneys, up-slanted eyes, an enlarged tongue, and abnormal dermal ridge patterns on the fingers, palms, and soles. Other forms of aneuploidy in humans result from abnormal numbers of sex chromosomes. Turner syndrome is a condition in which females have only one X chromosome. Symptoms may include short stature, webbed neck, kidney or heart malformations, underdeveloped sex characteristics, or sterility. Klinefelter syndrome is a condition in which males have one extra female sex chromosome, resulting in an XXY pattern. (Other, less frequent, chromosomal patterns include XXXY, XXXXY, XXYY, and XXXYY.) Symptoms of Klinefelter syndrome may include sterility, a tall physique, lack of secondary sex characteristics, breast development, and learning disabilities.

Chromosomal Abnormalities in Humans

Chromosomes are thread-like structure present in the nucleus that carries hereditary information in the form of genes which is passed from parents to offspring. Every species has a characteristic structure and number of chromosome present. Due to certain irregularities at the time of cell division, alteration in the structure or number of chromosomes may happen. Even the slightest alteration can lead to various abnormalities.

Chromosomal Disorders in Humans

Each human cell contains 46 (2n) chromosomes present as 23 pairs (n pairs), out of which 22 are autosomes and 1 pair of sex chromosomes.

Chromosomal disorders result from structural changes or numerical changes in chromosomes.

Chromosomal disorders due to numerical abnormalities

Chromosomal disorders are due to the change in the number of chromosomes present. This can be categorised into various types:

- Aneuploidy: loss or gain of a chromosome. This happens due to non-disjunction of chromatids when chromatids fail to separate during cell division. This results in one gamete having two copies of one chromosome and the other having no chromosome.
    i. Trisomy: The cell has one extra chromosome (2n+1)
    ii. Monosomy: The cell has one chromosome less (2n-1)

Aneuploidy can be due to nondisjunction of autosomes i.e. chromosomes 1-22 or sex chromosomes.

Karyotype of Down's syndrome Fig. 7

Klinefelter's syndrome Fig. 8

Chromosomal disorders due to structural abnormalities

This happens when a large set of genes are deleted, duplicated or rearranged causing structural changes in the chromosome. Structural abnormalities can be due to:

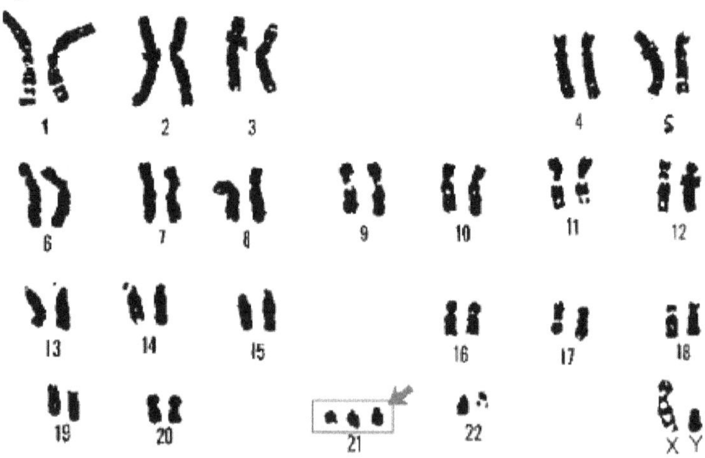

Fig. 7 Karyotype of Down's syndrome

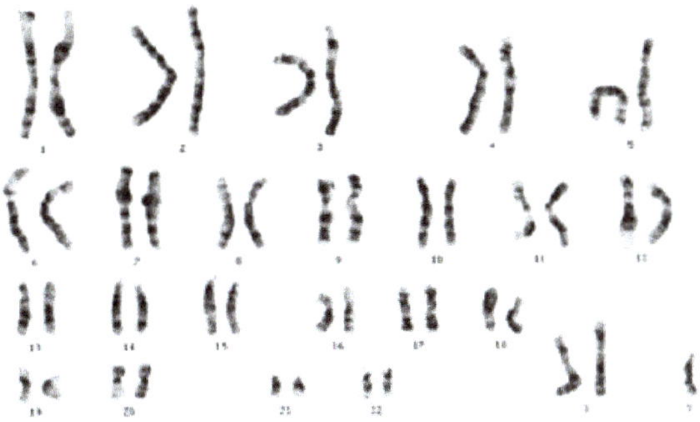

Fig. 8 Klinefelter's syndrome

1. Deletion: A portion of the chromosome is lost during cell division. A portion of chromosome without the centromere lags during anaphase movement and are lost from reorganising nuclei or digested by nucleases. The resulting chromosome lacks certain genes, that gets inherited to offspring. This condition is usually lethal due to missing genes.
   - Deletion can be terminal, where a terminal portion of a chromosome breaks resulting in one break
   - Intercalary deletion, where an intermediate portion is lost resulting from two breaks, which results in 3 pieces. The middle piece is lost and the other two parts rejoin

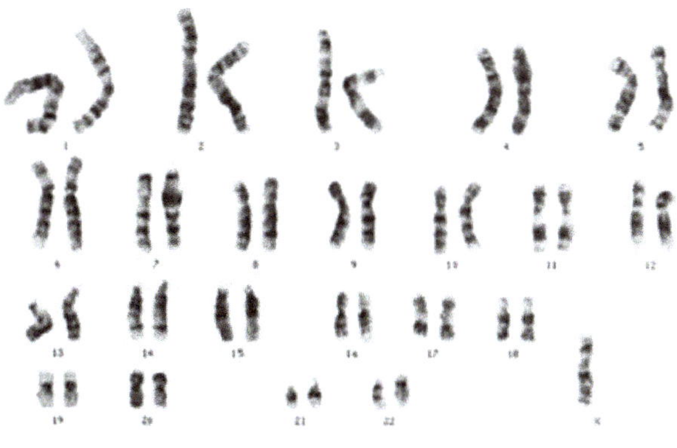

Fig. 9 Turner's syndrome

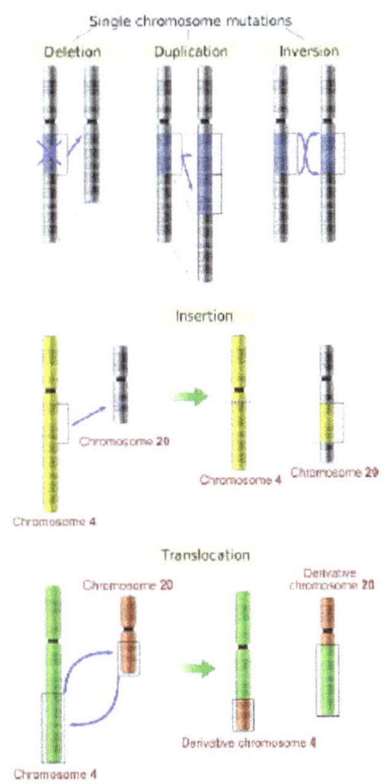

Fig. 10 Chromosomal disorders due to structural abnormalities

Chromosomal disorders due to aneuploidy: This is the cause of most of the genetically inherited disorders and abortion during pregnancy

Genetic Disorder	Defect in	Genotype	No. of Chromosomes	Phenotypic effect
Down's syndrome	Autosomes	Trisomy of 21st chromosome	47	• 1:700 live births
				• Short statured with a small round head
				• Furrowed tongue with partially open mouth
				• Broad palm with palm crease
				• Called as mongolism or Mongolian idiocy
				• Susceptible to heart disease and respiratory problems
				• Shorter life span
				• Physical, psychomotor and mental development is retarded
Patau syndrome		Trisomy of 13th chromosome	47	• 1:5000 live births
				• Rarely live for more than a few months
				• Sloping forehead, hair clip and cleft palate
				• Serious defects in the eyes, brain, kidney or circulatory system
				• Mentally retarded
Edward's syndrome		Trisomy of 18th chromosome	47	• 1:10000 live births
				• Multiple malformations like in lower jaw, fingers, skull, face and feet
				• Cardiac malformations
				• Can't live for more than 3-4 months
Klinefelter's syndrome	Sex chromosome	XXY	47	• These males have tall and masculine stature with feminine characteristics
				• Development of breasts (gynecomastia)
				• Sterile
				• Small testicles, high-pitched voice and sparse body hairs
				• Mental retardation
Turner's syndrome		Monosomy-XO	45	• 1:5000 live births
				• Only viable monosomy in humans
				• Sterile females with short stature
				• Breasts and ovaries are underdeveloped, thin and less pubic hairs
Triple X syndrome		XXX	47	• 1:1000 live births
				• They are called super females

				• Mild development delays and menstrual irregularities
XYY syndrome		XYY	47	• Males with an unusual height
				• Severe acne
				• Below normal intelligence

Example of disorder due to deletion:

Cri du chat (cry of the cat): Deletion of a small portion of 5th chromosome. Children with this disease have a small head with unusual facial features, severe mental retardation and make a sound like a cat while crying.

2. Duplication: The presence of part of a chromosome in excess is known as duplication. If the duplication is present only in one of the homologous pair of a chromosome, the duplicated part makes a loop to maximise juxtaposition of homologous regions during pairing. The extra segment can be arranged in many ways:
    - Tandem duplication, where the duplicated region is present side by side (ABCDEF→ABCDEDEF)
    - Reverse tandem, here duplicated region is just reverse of the normal sequence (ABCDEF→ABCDEEDF)
    - Displaced duplication, here duplicated region is not situated adjacent to the normal sequence
    - Transposed duplication, duplicated part becomes attached to a non-homologous chromosome
    - Extra-chromosomal duplication, here duplicated part acts as an independent chromosome in the presence of centromere

    Example of disorder due to duplication:

    Fragile X: Affects 1:1500 males and 1:2500 females. This is the most common form of mental retardation, where the CGS segment is repeated more than 200 times.

3. **Inversion:** inversion results from breakage and reunion of a part of the chromosome rotating by 180° on its own axis. So there occurs a rearrangement of genes. Its effects are not as severe as in other structural defects
4. **Translocation:** The shifting or transfer of a set of genes or part of a chromosome to a non-homologous one is known as translocation. There is no addition or loss of genes, only the rearrangement occurs. This rearrangement may lead to phenotype changes pertaining to the new environment. It can cause difficulties in the development of egg, sperm or zygote. These often result in miscarriages and children born with disabilities.

- Reciprocal translocation, here segment of two chromosomes gets interchanged
- Robertsonian translocation, here an entire chromosome attaches to another chromosome

  Example of disorder due to translocation:

  Acute MyelogenousLeukemia: In this type of cancer, bone marrow and cells derived from it show the presence of a short chromosome named as "Philadelphia (Ph1) chromosome". The 22nd chromosome loses a part of its arm which gets translocated to the distal end of the 9th chromosome. It is not transmitted to the offspring.

4

MUTATION

Mutation Definition

"Mutation is the change in our DNA base pair sequence due to various environmental factors such as UV light, or mistakes during DNA replication."

The DNA sequence is specific to each organism. It can sometimes undergo changes in its base-pairs sequence. It is termed as a mutation. A mutation may lead to changes in proteins translated by the DNA. Usually, the cells can recognize any damage caused by mutation and repair it before it becomes permanent.

TYPES OF MUTATIONS

The mutations are of three types:

- **Silent mutation:** It refers to any change in the sequence of DNA, which has no further impact on the amino acid sequence in a protein or in the functions performed by a protein. There is no phenotypic indicator of mutation.
- **Nonsense mutation:** When due to a point mutation there is change in the sequence of base pair leading to stop codon. It results in a shortened or non-functional protein.
- **Missense mutation:** A missense mutation is a result of point mutation leading to change in the codon, which then codes for another amino acid.

CAUSES OF MUTATIONS

The mutation leads to genetic variations among species. Positive mutations are transferred to successive generations, e.g., Mutation in the gene coding for haemoglobin causes sickle cell anaemia. The R.B.Cs become sickle in shape. However, in the African population, this mutation provides protection against malaria.

A mutation in the gene controlling the cell division leads to cancer.

OVERVIEW OF THE CAUSES AND IMPACTS OF MUTATION

Let us have an overview of the causes and impacts of mutation.

The mutation is caused due to the following reasons:

Internal Causes

Most of the mutations occur when the DNA fails to copy accurately. All these mutations lead to evolution. During cell division, the DNA makes a copy of its own. Sometimes, the copy of the DNA is not perfect and this slight difference from the original DNA is called a mutation.

External Causes

When the DNA is exposed to certain chemicals or radiations, it causes the DNA to break down. The ultraviolet radiations cause the thymine dimers to break resulting in a mutated DNA.

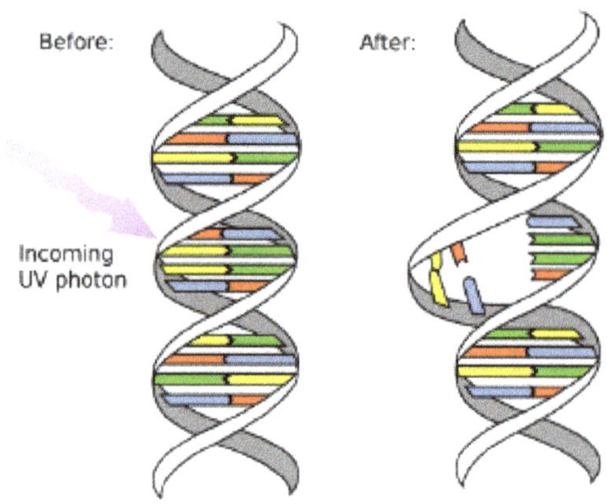

Fig. 1 External Causes

EFFECTS OF MUTATION

There are several mutations that cannot be passed on to the offsprings. Such mutations occur in the somatic cells and are known as somatic mutations.

The germline mutations can be passed on to successive generations and occur in the reproductive cells.

Let us have a look at some of the effects of mutation:

Beneficial Effects of Mutation

1. Few mutations result in new versions of proteins and help the organisms to adapt to changes in the environment. Such mutations lead to evolution.
2. Mutations in many bacteria result in antibiotic-resistant strains of bacteria that can survive in the presence of antibiotics.
3. A unique mutation found in the population of Italy protects them from atherosclerosis, where fatty materials build up in the blood vessels.

Effects of Mutations

1. Genetic disorders can be caused by the mutation of one or more genes. Cystic fibrosis is one such genetic disorder caused by the mutation in one or more genes.
2. Cancer is another disease caused by the mutation in genes that regulate the cell cycle.

DNA MUTATIONS

Recognize the impact of DNA mutations

Over a lifetime, our DNA can undergo changes or mutations in the sequence of bases: A, C, G and T. This results in changes in the proteins that are made. This can be a bad or a good thing.

A mutation is a change that occurs in our DNA sequence, either due to mistakes when the DNA is copied or as the result of environmental factors such as UV light and cigarette smoke. Mutations can occur during DNA replication if errors are made and not corrected in time. Mutations can also occur as the result of exposure to environmental factors such as smoking, sunlight and radiation. Often cells can recognize any potentially mutation-causing damage and repair it before it becomes a fixed mutation.

Mutations contribute to genetic variation within species. Mutations can also be inherited, particularly if they have a positive effect. For example, the disorder sickle cell anaemia is caused by a mutation in the gene that instructs the building of a protein called hemoglobin. This causes the red blood cells to become an abnormal, rigid, sickle shape. However, in African populations, having this mutation also protects against malaria.

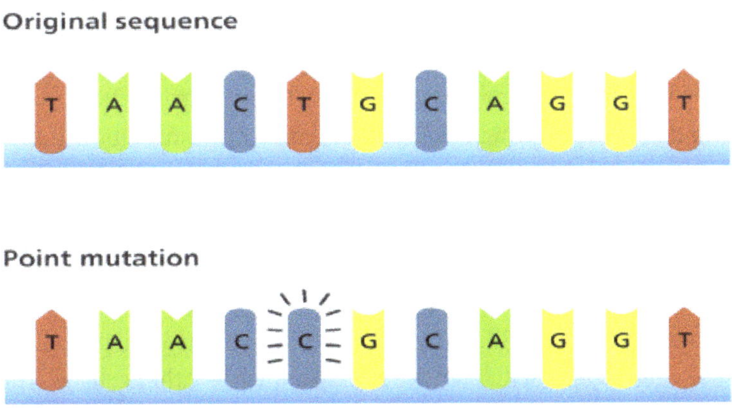

Figure 2. An illustration to show an example of a DNA mutation

However, mutation can also disrupt normal gene activity and cause diseases, like cancer. Cancer is the most common human genetic disease; it is caused by mutations occurring in a number of growth-controlling genes. Sometimes faulty, cancer-causing genes can exist from birth, increasing a person's chance of getting cancer.

MUTATIONS IN SOMATIC CELLS AND IN GAMETES

A gene mutation is a permanent alteration in the DNA sequence that makes up a gene, such that the sequence differs from what is found in most people. Mutations range in size; they can affect anywhere from a single DNA building block (base pair) to a large segment of a chromosome that includes multiple genes.

Gene mutations can be classified in two major ways:

- Hereditary mutations are inherited from a parent and are present throughout a person's life in virtually every cell in the body. These mutations are also called germline mutations because they are present in the parent's egg or sperm cells, which are also called germ cells. When an egg and a sperm cell unite, the resulting fertilized egg cell receives DNA from both parents. If this DNA has a mutation, the child that grows from the fertilized egg will have the mutation in each of his or her cells.
- Acquired (or somatic) mutations occur at some time during a person's life and are present only in certain cells, not in every cell in the body. These changes can be caused by environmental factors such as ultraviolet radiation from the sun, or can occur if a mistake is made as DNA copies itself during cell division. Acquired mutations in somatic cells (cells other than sperm and egg cells) cannot be passed on to the next generation.

Genetic changes that are described as de novo (new) mutations can be either hereditary or somatic. In some cases, the mutation occurs in a person's egg or sperm cell but is not present in any of the person's other cells. In other cases, the mutation occurs in the fertilized egg shortly after the egg and sperm cells unite. (It is often impossible to tell exactly when a de novo mutation happened.) As the fertilized egg divides, each resulting cell in the growing embryo will have the mutation. De novo mutations may explain genetic disorders in which an affected child has a mutation in every cell in the body but the parents do not, and there is no family history of the disorder.

Somatic mutations that happen in a single cell early in embryonic development can lead to a situation called mosaicism. These genetic changes are not present in a parent's egg or sperm cells, or in the fertilized egg, but happen a bit later when the embryo includes several cells. As all the cells divide during growth and development, cells that arise from the cell with the altered gene will have the mutation, while other cells will not. Depending on the mutation and how many cells are affected, mosaicism may or may not cause health problems.

Most disease-causing gene mutations are uncommon in the general population. However, other genetic changes occur more frequently. Genetic alterations that occur in more than 1 percent of the population are called polymorphisms. They are common enough to be considered a normal variation in the DNA. Polymorphisms are responsible for many of the normal differences between people such as eye color, hair color, and blood type. Although many polymorphisms have no negative effects on a person's health, some of these variations may influence the risk of developing certain disorders.

MAJOR TYPES OF MUTATIONS

So, how do mutations occur? The answer to this question is closely linked to the molecular details of how both DNA and the entire genome are organized. The smallest mutations are point mutations, in which only a single base pair is changed into another base pair. Yet another type of mutation is the nonsynonymous mutation, in which an amino acid sequence is changed. Such mutations lead to either the production of a different protein or the premature termination of a protein.

As opposed to nonsynonymous mutations, synonymous mutations do not change an amino acid sequence, although they occur, by definition, only in sequences that code for amino acids. Synonymous mutations exist because many amino acids are encoded by multiple codons. Base pairs can also have diverse regulating properties if they are located in introns, intergenic regions, or even within the coding sequence of genes. For some historic reasons, all of these groups are often subsumed with synonymous mutations under the label "silent" mutations. Depending on their function, such silent mutations can be anything from truly silent to extraordinarily important, the latter implying that working sequences are kept constant by purifying selection. This is the most likely explanation for the existence of ultraconserved noncoding elements that have survived for more than 100 million years without substantial change, as found by comparing the genomes of several vertebrates (Sandelin *et al.*, 2004).

Mutations may also take the form of insertions or deletions, which are together known as indels. Indels can have a wide variety of lengths. At the short end of the spectrum, indels of one or two base pairs within coding sequences have the greatest effect, because they will inevitably cause a frameshift (only the addition of one or more three-base-pair codons will keep a protein approximately intact). At the intermediate level, indels can affect parts of a gene or whole groups of genes. At the largest level, whole chromosomes or even whole copies of the genome can be affected by insertions or deletions, although such mutations are usually no longer subsumed under the label indel. At this high level, it is also possible to invert or translocate entire sections of a chromosome, and chromosomes can even fuse or break apart. If a large number of genes are lost as a result of one of these processes, then the consequences are usually very harmful. Of course, different genetic systems react differently to such events.

Finally, still other sources of mutations are the many different types of transposable elements, which are small entities of DNA that possess a mechanism that permits them to move around within the genome. Some of these elements copy and paste themselves into new locations, while others use a cut-and-paste method. Such movements can disrupt existing gene functions (by insertion in the middle of another gene), activate dormant gene functions (by perfect excision from a gene that was switched off by an earlier insertion), or occasionally lead to the production of new genes (by pasting material from different genes together).

A well-studied example of a mutation is seen in people suffering fromxerodermapigmentosa. Affected individuals have skin that is highly sensitive to UV rays from the sun.

When individuals are exposed to UV, pyrimidine dimers, especially those of thymine, are formed; people with xerodermapigmentosa are not able to repair the damage. These are not repaired because of a defect in the nucleotide excision repair enzymes, whereas in normal individuals, the thymine dimers are excised and the defect is corrected. The thymine dimers distort the structure of the DNA double helix, and this may cause problems during DNA replication. People with xerodermapigmentosa may have a higher risk of contracting skin cancer than those who don't have the condition.

Errors during DNA replication are not the only reason why mutations arise in DNA. Mutations, variations in the nucleotide sequence of a genome, can also occur because of damage to DNA. Such mutations may be of two types: induced or spontaneous. Induced mutations are those that result from an exposure to chemicals, UV rays, x-rays, or some other environmental agent. Spontaneous mutations occur without any exposure to any environmental agent; they are a result of natural reactions taking place within the body.

Mutations may have a wide range of effects. Some mutations are not expressed; these are known as silent mutations. Point mutations are those mutations that affect a single base pair. The most common nucleotide mutations are substitutions, in which one base is replaced by another. These can be of two types, either transitions or transversions. Transition substitution refers to a purine or pyrimidine being replaced by a base of the same kind; for example, a purine such as adenine may be replaced by the purine guanine. Transversion substitution refers to a purine being replaced by a pyrimidine, or vice versa; for example, cytosine, a pyrimidine, is replaced by adenine, a purine. Mutations can also be the result of the addition of a base, known as an insertion, or the removal of a base, also known as deletion. Sometimes a piece of DNA from one chromosome may get moved to another chromosome or to another region of the same chromosome; this is also known as translocation.

Mutations in repair genes have been known to cause cancer. Many mutated repair genes have been implicated in certain forms of pancreatic cancer, colon cancer, and colorectal cancer. Mutations can affect either somatic cells or germ cells. If many mutations accumulate in a somatic cell, they may lead to problems such as the uncontrolled cell division observed in cancer. If a mutation takes place in germ cells, the mutation will be passed on to the next generation, as in the case of hemophilia and xerodermapigmentosa.

EFFECTS OF MUTATIONS

A single mutation can have a large effect, but in many cases, evolutionary change is based on the accumulation of many mutations with small effects. Mutational effects can be beneficial, harmful, or neutral, depending on their context or location.

Most non-neutral mutations are deleterious. In general, the more base pairs that are affected by a mutation, the larger the effect of the mutation, and the larger the mutation's probability of being deleterious.

To better understand the impact of mutations, researchers have started to estimate distributions of mutational effects (DMEs) that quantify how many mutations occur with what effect on a given property of a biological system. In evolutionary studies, the property of interest is fitness, but in molecular systems biology, other emerging properties might also be of interest.

It is extraordinarily difficult to obtain reliable information about DMEs, because the corresponding effects span many orders of magnitude, from lethal to neutral to advantageous; in addition, many confounding factors usually complicate these analyses. To make things even more difficult, many mutations also interact with each other to alter their effects; this phenomenon is referred to as epistasis.

However, despite all these uncertainties, recent work has repeatedly indicated that the overwhelming majority of mutations have very small effects (Figure 1; Eyre-Walker &Keightley, 2007). Of course, much more work is needed in order to obtain more detailed information about DMEs, which are a fundamental property that governs the evolution of every biological system.

THE CAUSES OF GENETIC MUTATIONS

Mutations have many possible causes. Some mutations seem to happen spontaneously without any outside influence. They can occur when mistakes are made during DNA replication or transcription. Other mutations are caused by environmental factors. Anything in the environment that can cause a mutation is known as a mutagen. Examples of mutagens are pictured in Fig. 3.

Examples of Mutagens. Types of mutagens include radiation, chemicals, and infectious agents. Do you know of other examples of each type of mutagen shown here?

THE CHERNOBYL DISASTER: FOLLOW-UP

Though the area immediately around the Chernobyl disaster may not be safe for human life for thousands of years, the "Exclusion Zone" around the Chernobyl nuclear power station has become a haven for wildlife. As humans were evacuated from the area after the disaster, existing animal populations multiplied and rare species not seen for centuries have returned or have been reintroduced, including the lynx, wild boar, wolf, Eurasian brown bear, European bison, Przewalski's horse, and eagle owl. The Exclusion Zone is so lush with wildlife and greenery that in 2007 the Ukrainian government designated it a wildlife sanctuary. It is now one of the largest wildlife sanctuaries in Europe.

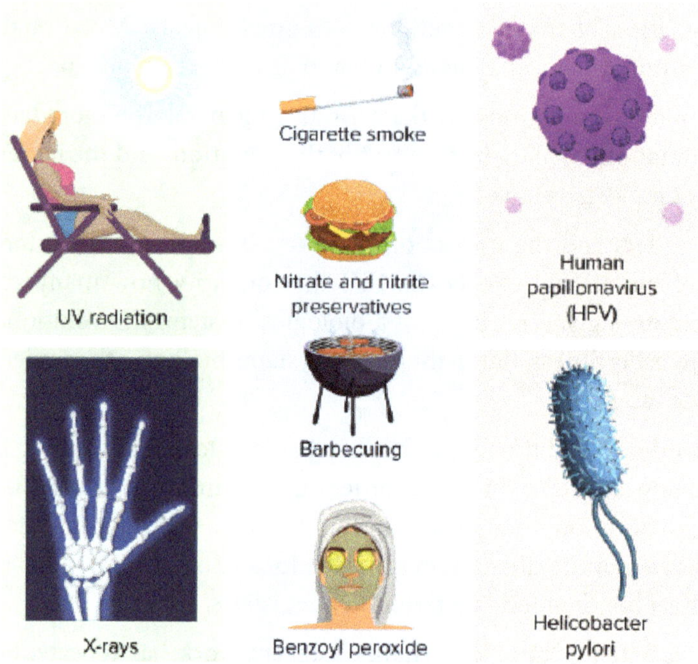

Fig. 3 The Causes of Genetic Mutations

SPONTANEOUS MUTATIONS

There are five common types of spontaneous mutations. These are described in the table below.

Mutation	Description
Tautomerism	a base is changed by the repositioning of a hydrogen atom
Depurination	loss of a purine base (A or G)
Deamination	spontaneous deamination of 5-methycytosine
Transition	a purine to purine (A to G, G to A), or a pyrimidine to pyrimidine (C to T, T to C) change
Transversion	a purine becomes a pyrimidine, or vice versa

5

NUCLEUS, NUCLEOLUS AND CHROMOSOMES

NUCLEUS

Typically, the nucleus is the most prominent organelle in a cell. The **nucleus** (plural = *nuclei*) houses the cell's DNA and directs the synthesis of ribosomes and proteins. Let's look at it in more detail (Fig. 1).

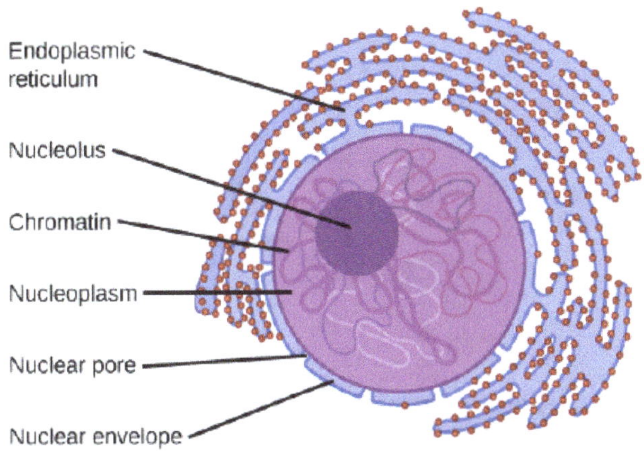

Fig. 1. The nucleus stores chromatin (DNA plus proteins) in a gel-like substance called the nucleoplasm. The nucleolus is a condensed region of chromatin where ribosome synthesis occurs. The boundary of the nucleus is called the nuclear envelope. It consists of two phospholipid bilayers: an outer membrane and an inner membrane. The nuclear membrane is continuous with the endoplasmic reticulum. Nuclear pores allow substances to enter and exit the nucleus.

Nuclear Envelope

The **nuclear envelope** is a double-membrane structure that constitutes the outermost portion of the nucleus (Figure 1). Both the inner and outer membranes of the nuclear envelope are phospholipid bilayers.

The nuclear envelope is punctuated with pores that control the passage of ions, molecules, and RNA between the nucleoplasm and cytoplasm. The **nucleoplasm** is the semi-solid fluid inside the nucleus, where we find the chromatin and the nucleolus.

CHROMATIN AND CHROMOSOMES

To understand chromatin, it is helpful to first consider chromosomes. **Chromosomes** are structures within the nucleus that are made up of DNA, the hereditary material.

In prokaryotes, DNA is organized into a single circular chromosome. In eukaryotes, chromosomes are linear structures. Every eukaryotic species has a specific number of chromosomes in the nuclei of its body's cells. For example, in humans, the chromosome number is 46, while in fruit flies, it is eight.

Chromosomes are only visible and distinguishable from one another when the cell is getting ready to divide. When the cell is in the growth and maintenance phases of its life cycle, proteins are attached to chromosomes, and they resemble an unwound, jumbled bunch of threads. These unwound protein-chromosome complexes are called **chromatin** (Fig. 2); chromatin describes the material that makes up the chromosomes both when condensed and decondensed.

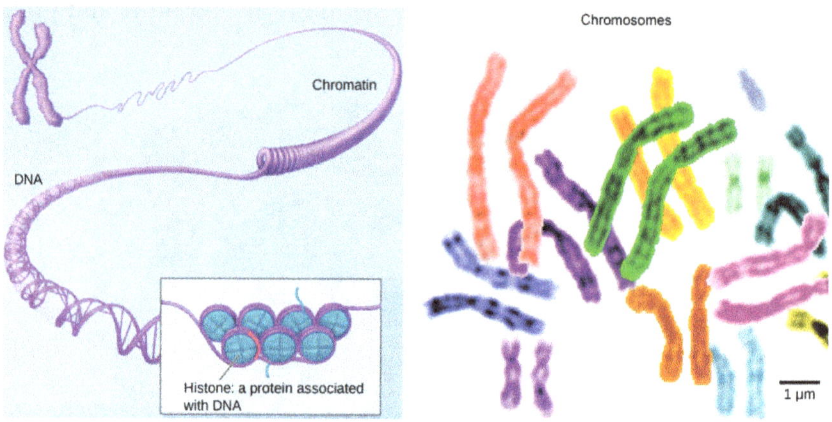

Fig. 2 Chromatin and Chromosomes

NUCLEOLUS

We already know that the nucleus directs the synthesis of ribosomes, but how does it do this? Some chromosomes have sections of DNA that encode ribosomal RNA. A darkly staining area within the nucleus called the nucleolus (plural = nucleoli) aggregates the ribosomal RNA with associated proteins to assemble the ribosomal subunits that are then transported out through the pores in the nuclear envelope to the cytoplasm.

CHROMOSOMES

Chromosome Meaning and Discovery

Chromosome means 'coloured body', that refers to its staining ability by certain dyes.

Karl Nägeli in 1842, first observed the rod-like structure present in the nucleus of the plant cell.

W. Waldeyer in 1888 coined the term 'chromosome'.

Walter Sutton and Theodor Boveri in 1902 suggested that chromosomes are the physical carrier of genes in the eukaryotic cells.

In the nucleus of each cell, the DNA molecule is packaged into thread-like structures called chromosomes. Each chromosome is made up of DNA tightly coiled many times around proteins called histones that support its structure.

Chromosomes are not visible in the cell's nucleus—not even under a microscope—when the cell is not dividing. However, the DNA that makes up chromosomes becomes more tightly packed during cell division and is then visible under a microscope. Most of what researchers know about chromosomes was learned by observing chromosomes during cell division.

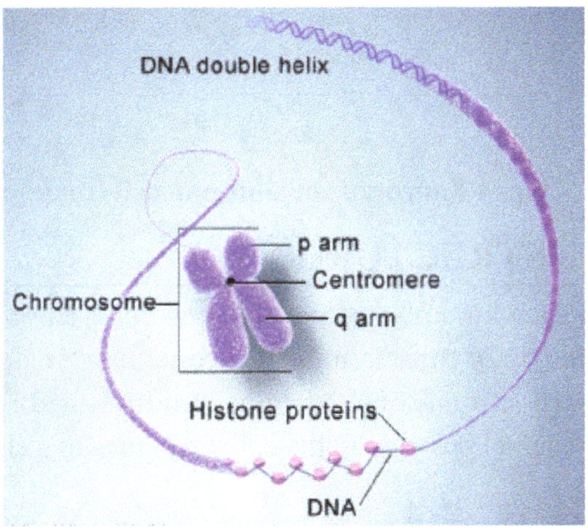

Fig. 3 DNA and histone proteins are packaged into structures called chromosomes.

Each chromosome has a constriction point called the centromere, which divides the chromosome into two sections, or "arms." The short arm of the chromosome is labeled the "p arm." The long arm of the chromosome is labeled the "q arm." The location of the centromere on each chromosome gives the chromosome its characteristic shape, and can be used to help describe the location of specific genes.

The number of chromosomes in any species is constant for all the cells. The number of chromosomes in gametes (e.g. sperms, egg) is half of the somatic cell and known as a haploid set of chromosomes, which is the result of meiosis during sexual reproduction. Chromosome number is preserved in the mitotic division of somatic cells, which is required for an organism to grow, repair and regenerate.

Chromosome number varies in different species. A nematode species contains only 2 chromosomes in a cell, whereas a protozoan species contains as much as 1600 chromosomes in the cell. Most of the plant and animal species contain 8 to 50 number of chromosomes in its somatic cell. The number of chromosomes does not reflect the complexity of a species. A human cell contains total 23 pair of chromosomes (2n, total 23×2=46), of which 22 are autosomes and 1 sex chromosome.

Karyotyping is a technique to study the structure of chromosomes present in a species. Chromosomes are isolated, stained and photographed. This technique is useful in finding out any chromosomal abnormalities.

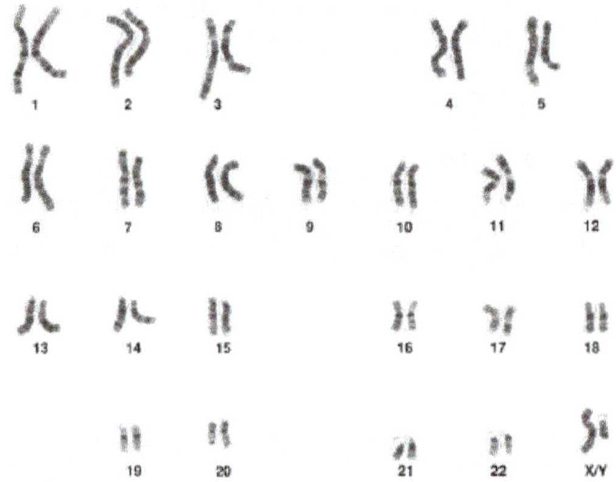

Fig. 4 Karyotype of a human cell (male)

CHROMOSOME STRUCTURE

Each cell has a pair of each kind of chromosome known as a homologous chromosome. Chromosomes are made up of chromatin, which contains a single molecule of DNA and associated proteins. Each chromosome contains hundreds and thousands of genes that can precisely code for several proteins in the cell. Structure of a chromosome can be best seen during cell division.

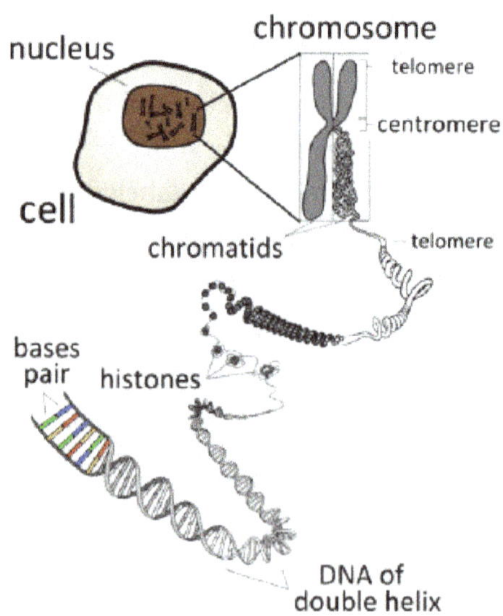

Fig. 5 Chromosome Structure

Main parts of chromosomes are:

- Chromatid: Each chromosome has two symmetrical structures called chromatids or sister chromatids which is visible in mitotic metaphase.
- Each chromatid contains a single DNA molecule
- At the anaphase of mitotic cell division, sister chromatids separate and migrate to opposite poles

Centromere and kinetochore: Sister chromatids are joined by the centromere.

- Spindle fibres during cell division are attached at the centromere
- The number and position of the centromere differs in different chromosomes
- The centromere is called primary constriction
- Centromere divides the chromosome into two parts, the shorter arm is known as 'p' arm and the longer arm is known as 'q' arm.
- The centromere contains a disc-shaped kinetochore, which has specific DNA sequence with special proteins bound to them
- The kinetochore provides the centre for polymerisation of tubulin proteins and assembly of microtubules

Secondary constriction and nucleolar organisers: Other than centromere, chromosomes possess secondary constrictions.

- Secondary constrictions can be identified from centromere at anaphase because there is bending only at the centromere (primary constriction)
- Secondary constrictions, which contain genes to form nucleoli are known as the nucleolar organiser

Telomere: Terminal part of a chromosome is known as a telomere.

- Telomeres are polar, which prevents the fusion of chromosomal segments

Satellite: It is an elongated segment that is sometimes present on a chromosome at the secondary constriction.

- The chromosomes with satellite are known as sat-chromosome

Chromatin: Chromosome is made up of chromatin. Chromatin is made up of DNA, RNA and proteins. At interphase, chromosomes are visible as thin chromatin fibres present in the nucleoplasm. During cell division, the chromatin fibres condense and chromosomes are visible with distinct features.

- The darkly stained, condensed region of chromatin is known as heterochromatin. It contains tightly packed DNA, which is genetically inactive
- The light stained, diffused region of chromatin is known as euchromatin. It contains genetically active and loosely packed DNA

- At prophase, the chromosomal material is visible as thin filaments known as chromonemata
- At interphase, bead-like structures are visible, which are an accumulation of chromatin material called chromomere. Chromatin with chromomere looks like a necklace with beads

Structural Organisation of Chromatin

- Chromatin consists of DNA and associated proteins. DNA is packaged in a highly organised manner in chromosomes
    1. Nucleosomes are the basic unit of chromatin. It is 10 nm in the diameter
        - DNA packing is facilitated by proteins called histones. DNA is wound around histone proteins to form a nucleosome
        - There are 5 types of histone proteins in the eukaryotic chromosomes, namely H1, H2A, H2B, H3 and H4
        - Histones are positively charged due to the presence of amino acids with basic side chains and it associates with negatively charged DNA due to the presence of phosphate groups
        - Histone proteins play an important role in gene regulation
        - A typical nucleosome contains 200 bp of DNA helix. The core particle of the nucleosome consists of approximately 146 base pairs (bp) of DNA coiled around a core of eight histone molecules (2 molecules of 4 histone proteins). That is linked by linker DNA of about 80 bp.
        - Nucleosomes prevent DNA from getting tangled

Linker DNA and the fifth histone (H1) pack adjacent nucleosomes to a 30 nm compact chromatin fibre

These fibres form a large coiled loop held together by non-histone proteins (actin, α and β tubulin, myosin) called scaffolding proteins to form extended chromatin which is 300 nm in diameter

Chromatin further condenses with the help of protein known as condensin, it binds to DNA and wraps it into coiled loops and we get the compacted chromosome

Types of Chromosome

Based on the number of centromeres present

i. **Monocentric:** having only one centromere

ii. **Holocentric:** having diffused centromere and microtubules are attached along the length of a chromosome

iii. **Acentric:** chromosome may break and fuse together to form a chromosome without a centromere. It cannot attach to the mitotic spindle

iv. **Dicentric:** chromosomal aberration where chromosomes break and fuse together with two centromeres. They are also unstable as two centromeres tend to migrate to opposite poles resulting in fragmentation

Based on the position of the centromere

v. **Telocentric:** rod-like chromosome with centromere present on the proximal end. No 'p' arm (short arm) present. Telomeric chromosomes are not found in humans

vi. **Acrocentric:** rod-like, centromere present at one end giving rise to one very short arm and an exceptionally long arm

vii. **Submetacentric:** L-shaped or J-shaped, with centromere near the centre of the chromosome giving rise to two unequal arms

viii. **Metacentric:** V-shaped chromosomes with centromere in the middle giving rise to two equal arms

GIANT CHROMOSOMES

i. **Polytene chromosome**

a. Balbiani first discovered a structure in the nuclei of secretory glands of midges
b. Painter, Heitz and Bauer, rediscovered them in the salivary gland of Drosophila and recognised them as a chromosome
c. Also known as Salivary gland chromosome
d. These are called polytene by Kollar due to the presence of many chromonemata in them
e. These are present in some cells of the larvae of Dipteran insects
f. These are very large due to the presence of high DNA content
g. The polytene chromosome of Drosophila's salivary gland has 1000 DNA molecules Chironomus has 1600 DNA molecules in its each polytene chromosome
h. There is a series of alternating dark and clear bands called interband
i. Chromosome puffs or Balbiani rings are present, which are the swelling of bands due to DNA unfolding into open loops. These are the region of the intense transcription or mRNA formation

ii. **Lampbrush chromosome**

a. First discovered in the oocytes of salamander
b. The name is given due to its resemblance with a brush that is used for cleaning lamp, glass chimneys, etc.
c. They occur at the diplotene stage of oocytes in vertebrates and invertebrates
d. Lampbrush chromosomes are also found in the spermatocytes of many animals and also found in the giant nucleus of an algae Acetabularia
e. They are present as a bivalent with 4 chromatids
f. Chromosomal axis is formed from highly condensed chromatin and lateral loops extend from the row of chromomeres

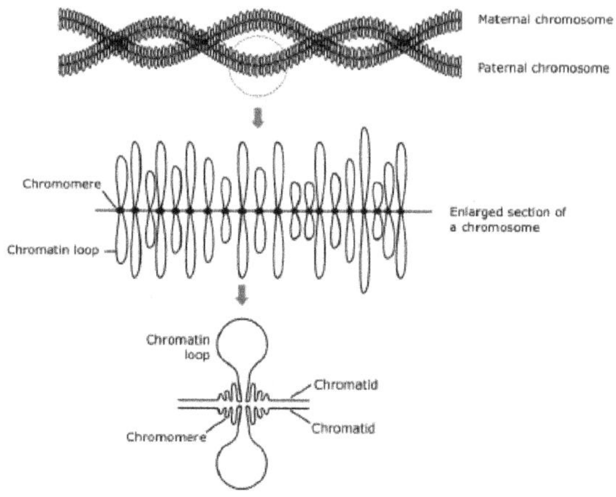

Fig. 6 Lampbrush chromosome

g. Lateral loops of DNA are always symmetrical and formed due to intense RNA synthesis
h. In the oocytes of salamander, there are 10,000 loops present per haploid set of chromosomes
i. The centromere doesn't bear any loops

PROPERTIES OF CHROMOSOMES

Humans have 23 pairs of chromosomes. This unique structure of the chromosome keeps DNA tightly packed with histone proteins to fit inside the cell and it also helps in giving more stability to it. For instance, the unwound DNA of a single cell will measure 6 feet, with this, one can understand the need for packaging inside the nucleus of each cell.

Cells must continuously repair, grow and regenerate to replace the old cells. Cell division is important for the growth and development of an organism. We can see chromosomes clearly during cell division. Chromosomes ensure that DNA is divided equally between the daughter cells during cell division. Even a small irregularity in the process may lead to various diseases and deformities. Uncontrolled cell division results in tumour cells and causes cancer. Chromosomal aberration like change in structure or number can cause genetic disorders, e.g. Down's syndrome, Turner's syndrome, etc. Defective chromosomes may even lead to a certain type of leukaemia in humans.

FUNCTIONS OF CHROMOSOMES

- The main function of chromosomes is to carry the genetic material from one generation to another
- Chromosomes play an important role and act as a guiding force in the growth, reproduction, repair and regeneration process, which is important for their survival
- Chromosomes protect the DNA from getting tangled and damaged

- Histone and non-histone proteins help in the regulation of gene expression
- Spindle fibres attached to the centromere help in the movement of the chromosome during cell division
- Each chromosome contains thousands of genes that precisely code for multiple proteins present in the body

6

INTRACELLULAR COMPARTMENTS

Cellular compartments in cell biology comprise all of the closed parts within the cytosol of a eukaryotic cell, usually surrounded by a single or double lipid layer membrane. These compartments are often, but not always, defined as membrane-bound organelles. The formation of cellular compartments is called compartmentalization.

Both organelles, the mitochondria and chloroplasts (in photosynthetic organisms), are compartments that are believed to be of endosymbiotic origin. Other compartments such as peroxisomes, lysosomes, the endoplasmic reticulum, the cell nucleus or the Golgi apparatus are not of endosymbiotic origin. Smaller elements like vesicles, and sometimes even microtubules can also be counted as compartments.

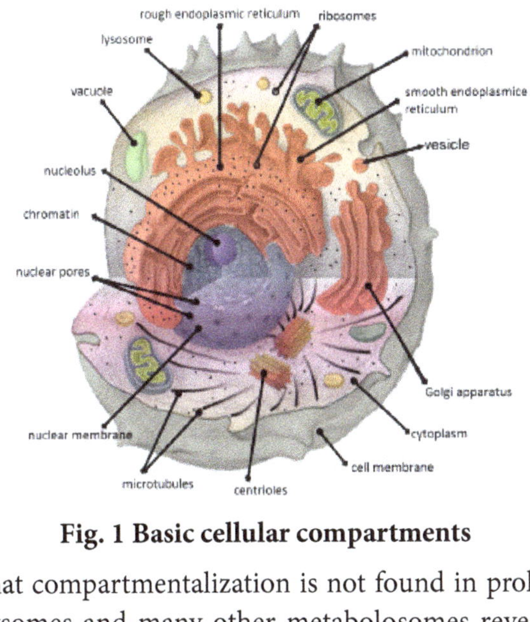

Fig. 1 Basic cellular compartments

It was thought that compartmentalization is not found in prokaryotic cells, but the discovery of carboxysomes and many other metabolosomes revealed that prokaryotic cells are capable of making compartmentalized structures, albeit these are in most cases not surrounded by a lipid bilayer, but of pure proteinaceous built.

CELLULAR COMPARTMENTS AND TRAFFIC

Introduction

This section reminds you of the numerous specialised intracellular compartments of the eukaryotic cell, with how molecules are moved rapidly and specifically between them in eukaryotic cells.

On average, these compartments occupy about half of the volume of the cell. Their associated membranes in a liver cell (hepatocyte), for example, constitute about 50 times more of the total area than the plasma membrane (Table 1). The extent of the intracellular compartments varies between cell types, but the contemporary picture of a cell is a labyrinth of membrane-bound intracellular compartments, endosomes and organelles, with a highly developed interconnecting system of transport vesicles.

Table 1 Relative amounts of membrane in two eukaryotic cells (% of total membrane, by surface area).

Membrane type	Hepatocyte	Pancreatic exocrine cell
Plasma membrane	2%	5%
Rough ER	35%	60%
Smooth ER	16%	<1%
Golgi apparatus	7%	10%
Mitochondrialinnermembrane	32%	17%
Mitochondrialoutermembrane	7%	4%
Nuclear membrane	0.2%	0.7%
Lysosomes, peroxisomes, endosomes, secretory vesicles	<2%	<4%

The volume of the hepatocyte is approximately 5000 μm3 with 110,000 μm^2 of membrane surface; the volume of the pancreatic cell is 1000 μm3 with an estimated 13,000 μm^2 of membrane surface. Note that the large amount of rough endoplasmic reticulum (ER) in the pancreatic cell reflects its function of secreting digestive enzymes; the relatively large number of mitochondria in liver cells relates to their high metabolic activity.

Vesicular transport is the mechanism by which molecules are ferried in membrane-bound vesicles between the membrane-bound compartments listed above, or between these compartments and the plasma membrane. Small transport vesicles pinch off from one compartment, and diffuse, or are more often actively transported, to another compartment, where they fuse and discharge their cargo. The size and shape of the vesicles, their cargo, their packaging and their means of transportation vary depending on the trafficking pathway involved. The cargo itself may be soluble within the compartment (e.g. a secreted hormone) or may be associated with membrane proteins. Topologically, the extracellular space, the ER, the Golgi network, the endosomes, secretory vesicles and transport vesicles are all equivalent spaces, each separated from the cytosol by one phospholipid bilayer. For molecules to move from the cytosol into one of these compartments they have to cross one membrane. In this chapter we are concerned only with transport between these topologically equivalent compartments mediated by the budding and fusion of vesicles.

Transport of proteins to the correct cellular compartment depends on signal sequences in the polypeptide chain.

Signal sequences at the N-terminus of newly synthesised polypeptides are recognised by a signal recognition protein, which directs them to sites on the ER, where they are transported into the lumen of the ER as they are translated. The signal sequence is enzymatically removed from most proteins after translation. For multipass membrane proteins, internal stop-transfer and start-transfer signals are required to produce the correct looping across the membrane.

Proteins destined for specific cellular compartments also contain signal sequences, although these may be non-contiguous within the polypeptide chain, and are referred to as signal patches. Unlike the N-terminal signal sequences, signal patches are integral to the final protein and are not cleaved off. The signal patches serve to sort the proteins for their appropriate compartment and/or ensure that they remain in that compartment. Proteins called sorting receptors recognise the signal patches and group the proteins in appropriate vesicles for transpo

Body Fluids and Fluid Compartments

he chemical reactions of life take place in aqueous solutions. The dissolved substances in a solution are called solutes. In the human body, solutes vary in different parts of the body, but may include proteins—including those that transport lipids, carbohydrates, and, very importantly, electrolytes. Often in medicine, a mineral dissociated from a salt that carries an electrical charge (an ion) is called and electrolyte. For instance, sodium ions (Na^+) and chloride ions (Cl^-) are often referred to as electrolytes.

In the body, water moves through semi-permeable membranes of cells and from one compartment of the body to another by a process called osmosis. Osmosis is basically the diffusion of water from regions of higher concentration to regions of lower concentration, along an osmotic gradient across a semi-permeable membrane. As a result, water will move into and out of cells and tissues, depending on the relative concentrations of the water and solutes found there. An appropriate balance of solutes inside and outside of cells must be maintained to ensure normal function.

Body Water Content

Human beings are mostly water, ranging from about 75 percent of body mass in infants to about 50–60 percent in adult men and women, to as low as 45 percent in old age. The percent of body water changes with develop-ment, because the proportions of the body given over to each organ and to muscles, fat, bone, and other tissues change from infancy to adulthood. Your brain and kidneys have the highest propor-tions of water, which composes 80–85 percent of their masses. In contrast, teeth have the lowest proportion of water, at 8–10 percent.

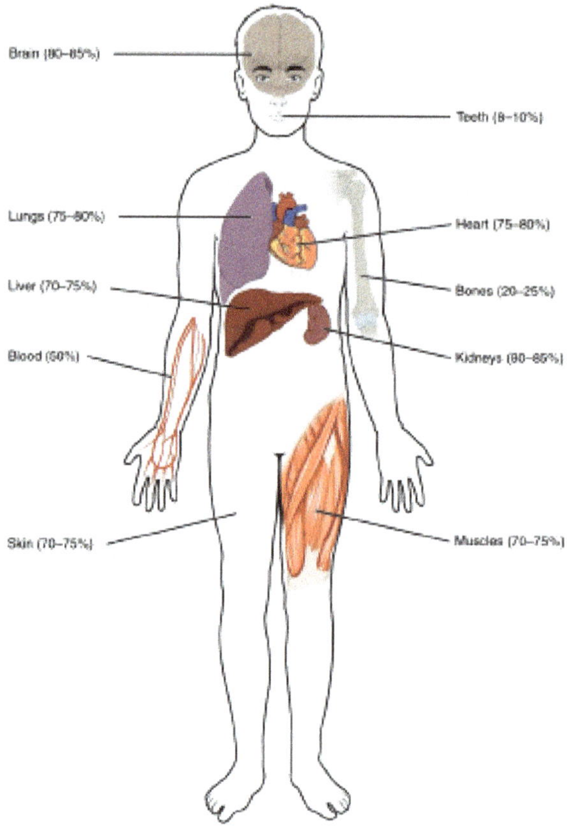

Fig. 2. Water content varies in different body organs and tissues, from as little as 8 percent in the teeth to as much as 85 percent in the brain.

Fluid Compartments

Body fluids can be discussed in terms of their specific fluid compartment, a location that is largely separate from another compartment by some form of a physical barrier. The intracellular fluid (ICF) compartment is the system that includes all fluid enclosed in cells by their plasma membranes. Extracellular fluid (ECF) surrounds all cells in the body. Extracellular fluid has two primary constituents: the fluid component of the blood (called plasma) and the interstitial fluid (IF) that surrounds all cells not in the blood.

Intracellular Fluid

The ICF lies within cells and is the principal component of the cytosol/cytoplasm. The ICF makes up about 60 percent of the total water in the human body, and in an average-size adult male, the ICF accounts for about 25 liters (seven gallons) of fluid. This fluid volume tends to be very stable, because the amount of water in living cells is closely regulated. If the amount of water inside a cell falls to a value that is too low, the cytosol becomes too concentrated with solutes to carry on normal cellular activities; if too much water enters a cell, the cell may burst and be destroyed.

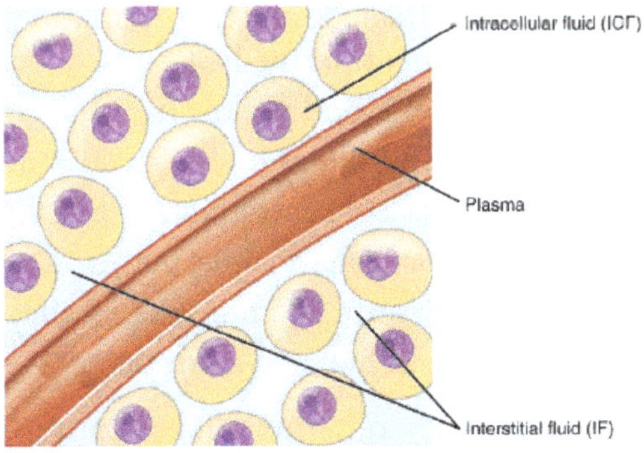

Fig. 3. The intracellular fluid (ICF) is the fluid within cells. The interstitial fluid (IF) is part of the extracellular fluid (ECF) between the cells. Blood plasma is the second part of the ECF. Materials travel between cells and the plasma in capillaries through the IF.

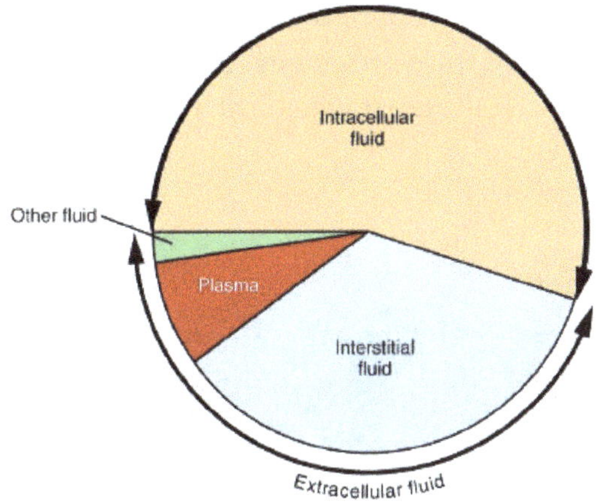

Fig. 4. Most of the water in the body is intracellular fluid. The second largest volume is the interstitial fluid, which surrounds cells that are not blood cells.

Extracellular Fluid

The ECF accounts for the other one-third of the body's water content. Approximately 20 percent of the ECF is found in plasma. Plasma travels through the body in blood vessels and transports a range of materials, including blood cells, proteins (including clotting factors and antibodies), electrolytes, nutrients, gases, and wastes. Gases, nutrients, and waste materials travel between capillaries and cells through the IF. Cells are separated from the IF by a selectively permeable cell membrane that helps regulate the passage of materials between the IF and the interior of the cell.

The body has other water-based ECF. These include the cerebrospinal fluid that bathes the brain and spinal cord, lymph, the synovial fluid in joints, the pleural fluid in the pleural cavities, the pericardial fluid in the cardiac sac, the peritoneal fluid in the peritoneal cavity, and the aqueous humor of the eye. Because these fluids are outside of cells, these fluids are also considered components of the ECF compartment.

COMPOSITION OF BODY FLUIDS

The compositions of the two components of the ECF—plasma and IF—are more similar to each other than either is to the ICF. Blood plasma has high concentrations of sodium, chloride, bicarbonate, and protein. The IF has high concentrations of sodium, chloride, and bicarbonate, but a relatively lower concentration of protein. In contrast, the ICF has elevated amounts of potassium, phosphate, magnesium, and protein. Overall, the ICF contains high concentrations of potassium and phosphate, whereas both plasma and the ECF contain high concentrations of sodium and chloride.

Most body fluids are neutral in charge. Thus, cations, or positively charged ions, and anions, or negatively charged ions, are balanced in fluids. As seen in the previous graph, sodium (Na^+) ions and chloride (Cl^-) ions are concentrated in the ECF of the body, whereas potassium (K^+) ions are concentrated inside cells. Although sodium and potassium can "leak" through "pores" into and out of cells, respectively, the high levels of potassium and low levels of sodium in the ICF are maintained by sodium-potassium pumps in the cell membranes. These pumps use the energy supplied by ATP to pump sodium out of the cell and potassium into the cell.

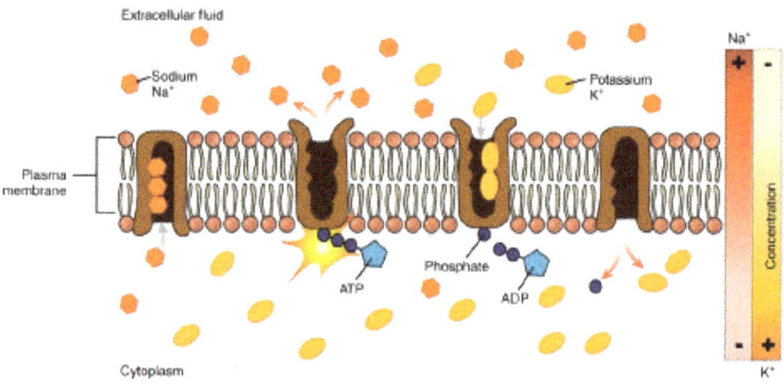

Fig. 5. The sodium-potassium pump is powered by ATP to transfer sodium out of the cytoplasm and into the ECF. The pump also transfers potassium out of the ECF and into the cytoplasm.

FLUID MOVEMENT BETWEEN COMPARTMENTS

Hydrostatic pressure, the force exerted by a fluid against a wall, causes movement of fluid between compartments. The hydrostatic pressure of blood is the pressure exerted by blood against the walls of the blood vessels by the pumping action of the heart. In capillaries, hydrostatic pressure (also known as capillary blood pressure) is higher than the opposing "colloid osmotic pressure" in blood—a "constant" pressure primarily produced by circulating albumin—at the arteriolar end of the capillary. This pressure forces plasma and nutrients out of the capillaries and into surrounding tissues. Fluid and the cellular wastes in the tissues enter the capillaries at the venule end, where the hydrostatic pressure is less than the osmotic pressure in the vessel. Filtration pressure squeezes fluid from the plasma in the blood to the IF surrounding the tissue cells. The surplus fluid in the interstitial space that is not returned directly back to the capillaries is drained from tissues by the lymphatic system, and then re-enters the vascular system at the subclavian veins.

Hydrostatic pressure is especially important in governing the movement of water in the nephrons of the kidneys to ensure proper filtering of the blood to form urine. As hydrostatic pressure in the kidneys increases, the amount of water leaving the capillaries also increases, and more urine filtrate is formed. If hydrostatic pressure in the kidneys drops too low, as can happen in dehydration, the functions of the kidneys will be impaired, and less nitrogenous wastes will be removed from the bloodstream. Extreme dehydration can result in kidney failure.

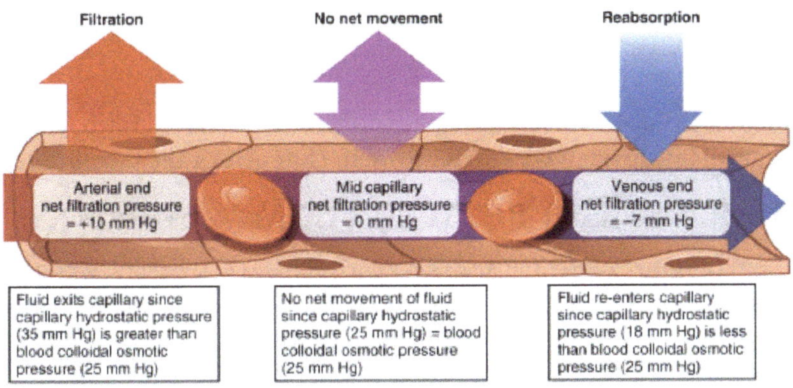

Fig. 6. Net filtration occurs near the arterial end of the capillary since capillary hydrostatic pressure (CHP) is greater than blood colloidal osmotic pressure (BCOP). There is no net movement of fluid near the midpoint of the capillary since CHP = BCOP. Net reabsorption occurs near the venous end of the capillary since BCOPis greater than CHP.

Fluid also moves between compartments along an osmotic gradient. Recall that an osmotic gradient is produced by the difference in concentration of all solutes on either side of a semi-permeable membrane. The magnitude of the osmotic gradient is proportional to the difference in the concentration of solutes on one side of the cell membrane to that on the other side. Water will move by osmosis from the side where its concentration is high (and the concentration of solute is low) to the side of the membrane where its concentration is low (and the concentration of solute is high). In the body, water moves by osmosis from plasma to the IF (and the reverse) and from the IF to the ICF (and the reverse). In the body, water moves constantly into and out of fluid compartments as conditions change in different parts of the body.

For example, if you are sweating, you will lose water through your skin. Sweating depletes your tissues of water and increases the solute concentration in those tissues. As this happens, water diffuses from your blood into sweat glands and surrounding skin tissues that have become dehydrated because of the osmotic gradient. Additionally, as water leaves the blood, it is replaced by the water in other tissues throughout your body that are not dehydrated. If this continues, dehydration spreads throughout the body. When a dehydrated person drinks water and rehydrates, the water is redistributed by the same gradient, but in the opposite direction, replenishing water in all of the tissues.

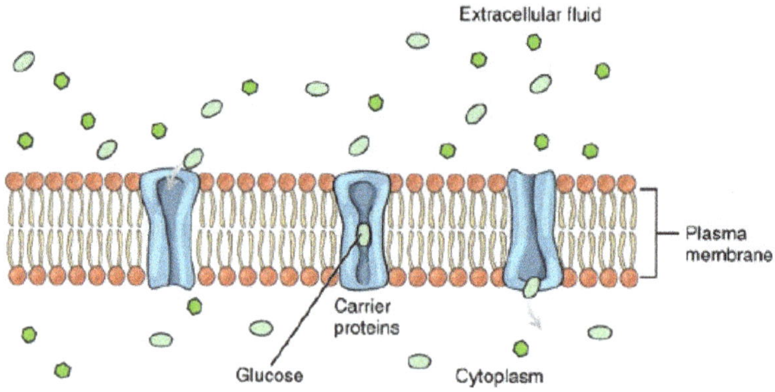

Fig. 7. Glucose molecules use facilitated diffusion to move down a concentration gradient through the carrier protein channels in the membrane

SOLUTE MOVEMENT BETWEEN COMPARTMENTS

The movement of some solutes between compartments is active, which consumes energy and is an active transport process, whereas the movement of other solutes is passive, which does not require energy. Active transport allows cells to move a specific substance against its concentration gradient through a membrane protein, requiring energy in the form of ATP. For example, the sodium-potassium pump employs active transport to pump sodium out of cells and potassium into cells, with both substances moving against their concentration gradients.

Passive transport of a molecule or ion depends on its ability to pass through the membrane, as well as the existence of a concentration gradient that allows the molecules to diffuse from an area of higher concentration to an area of lower concentration. Some molecules, like gases, lipids, and water itself (which also utilizes water channels in the membrane called aquaporins), slip fairly easily through the cell membrane; others, including polar molecules like glucose, amino acids, and ions do not. Some of these molecules enter and leave cells using facilitated transport, whereby the molecules move down a concentration gradient through specific protein channels in the membrane. This process does not require energy. For example, glucose is transferred into cells by glucose transporters that use facilitated transport.

7

THE MITOCHONDRION

Mitochondria Definitions

"Mitochondria are membrane-bound organelles present in the cytoplasm of all eukaryotic cells, that produces adinosine triphosphate (ATP), the main energy molecule used by the cell."

"Mitochondria are membrane-bound cell organelles (mitochondrion, singular) that generate most of the chemical energy needed to power the cell's biochemical reactions. Chemical energy produced by the mitochondria is stored in a small molecule called adenosine triphosphate (ATP). Mitochondria contain their own small chromosomes. Generally, mitochondria, and therefore mitochondrial DNA, are inherited only from the mother."

Popularly known as the "Powerhouse of the cell," mitochondria (singular: mitochondrion) are a double membrane-bound organelle found in most eukaryotic organisms. They are found inside the cytoplasm and essentially function as the cell's "digestive system."

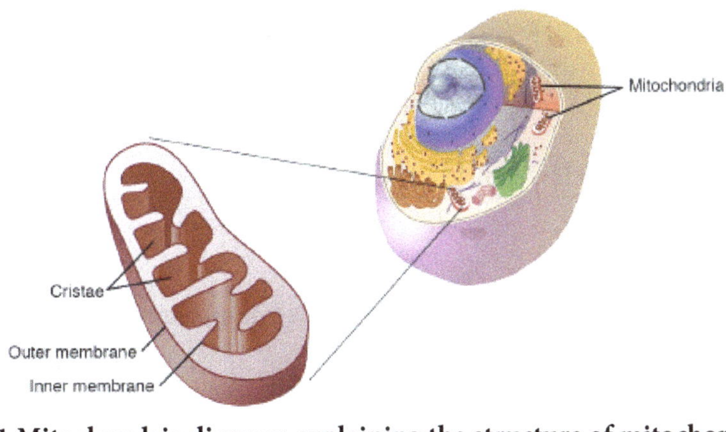

Fig. 1 Mitochondria diagram explaining the structure of mitochondria

They play a major role in breaking down nutrients and generating energy-rich molecules for the cell. Many of the biochemical reactions involved in cellular respiration take place within the mitochondria. The term 'mitochondrion' is derived from the Greek words "*mitos*" and "*chondrion*" which means "thread" and "granules-like", respectively. It was first described by a German pathologist named Richard Altmann in the year 1890.

Some of the **major regions** include the:

- **Outer membrane:** Small molecules can pass freely through the outer membrane. This outer portion includes proteins called porins, which form channels that allow proteins to cross. The outer membrane also hosts a number of enzymes with a wide variety of functions.
- **Intermembrane space:** This is the area between the inner and outer membranes.
- **Inner membrane:** This membrane holds proteins that have several roles. Because there are no porins in the inner membrane, it is impermeable to most molecules. Molecules can only cross the inner membrane in special membrane transporters. The inner membrane is where most ATP is created.
- **Cristae:** These are the folds of the inner membrane. They increase the surface area of the membrane, therefore increasing the space available for chemical reactions.
- **Matrix:** This is the space within the inner membrane. Containing hundreds of enzymes, it is important in the production of ATP. Mitochondrial DNA is housed here (see below).

Different cell types have different numbers of mitochondria. For instance, mature red blood cells have none at all, whereas liver cells can have more than 2,000. Cells with a high demand for energy tend to have greater numbers of mitochondria. Around 40 percent of the cytoplasm in heart muscle cells is taken up by mitochondria.

Although mitochondria are often drawn as oval-shaped organelles, they are constantly dividing (fission) and bonding together (fusion). So, in reality, these organelles are linked together in ever-changing networks.

Also, in sperm cells, the mitochondria are spiraled in the midpiece and provide energy for tail motion.

STRUCTURE OF MITOCHONDRIA

- The mitochondrion is a double-membraned, rod-shaped structure found in both plant and animal cell.
- Its size ranges from 0.5 to 1.0 micrometre in diameter.
- The structure comprises an outer membrane, an inner membrane, and a gel-like material called the matrix.

- The outer membrane and the inner membrane are made of proteins and phospholipid layers separated by the intermembrane space.
- The outer membrane covers the surface of the mitochondrion and has a large number of special proteins known as porins.
- It is freely permeable to ions, nutrient molecules, energy molecules like the ADP and ATP molecules.

Cristae

The inner membrane of mitochondria is rather complex in structure. It has many folds that form a layered structure called cristae, and this helps in increasing the surface area inside the organelle. The cristae and the proteins of the inner membrane aids in the production of ATP molecules. The inner membrane is strictly permeable only to oxygen and to ATP molecules. A number of chemical reactions take place within the inner membrane of mitochondria.

MITOCHONDRIAL MATRIX

The mitochondrial matrix is a viscous fluid that contains a mixture of enzymes and proteins. It also comprises ribosomes, inorganic ions, mitochondrial DNA, nucleotide cofactors, and organic molecules. The enzymes present in the matrix play an important role in the synthesis of ATP molecules.

FUNCTIONS OF MITOCHONDRIA

The most important function of mitochondria is to produce energy through the process of oxidative phosphorylation. It is also involved in the following process:

1. Regulates the metabolic activity of the cell
2. Promotes the growth of new cells and cell multiplication
3. Helps in detoxifying ammonia in the liver cells
4. Plays an important role in apoptosis or programmed cell death
5. Responsible for building certain parts of the blood and various hormones like testosterone and oestrogen
6. Helps in maintaining an adequate concentration of calcium ions within the compartments of the cell
7. It is also involved in various cellular activities like cellular differentiation, cell signalling, cell senescence, controlling the cell cycle and also in cell growth.

DISORDERS ASSOCIATED WITH MITOCHONDRIA

Any irregularity in the way mitochondria functions can directly affect human health, but often, it is difficult to identify because symptoms differ from person to person. Disorders of the mitochondria can be quite severe; in some cases, it can even cause an organ to fail.

MITOCHONDRIAL DNA

Although most of our DNA is kept in the nucleus of each cell, mitochondria have their own set of DNA. Interestingly, mitochondrial DNA (mtDNA) is more similar to bacterial DNA.

The mtDNA holds the instructions for a number of proteinsTrusted Source and other cellular support equipment across 37 genes.

The human genome stored in the nuclei of our cells contains around 3.3 billion base pairs, whereas mtDNA consists of less than 17,000Trusted Source.

During reproduction, half of a child's DNA comes from their father and half from their mother. However, the child always receives their mtDNA from their mother. Because of this, mtDNA has proven very useful for tracing genetic lines.

For instance, mtDNA analyses have concluded that humans may have originated in Africa relatively recently, around 200,000 years ago, descended from a common ancestor, known as mitochondrial EveTrusted Source.

WHAT DO MITOCHONDRIA DO?

Although the best-known role of mitochondria is energy production, they carry out other important tasks as well.

In fact, only about 3 percent of the genes needed to make a mitochondrion go into its energy production equipment. The vast majority are involved in other jobs that are specific to the cell type where they are found.

The roles of the mitochondria

Producing energy

ATP, a complex organic chemical found in all forms of life, is often referred to as the molecular unit of currency because it powers metabolic processes. Most ATP is produced in mitochondria through a series of reactions, known as the citric acid cycle or the Krebs cycle.

Energy production mostly takes place on the folds or cristae of the inner membrane.

Mitochondria convert chemical energy from the food we eat into an energy form that the cell can use. This process is called oxidative phosphorylation.

The Krebs cycle produces a chemical called NADH. NADH is used by enzymes embedded in the cristae to produce ATP. In molecules of ATP, energy is stored in the form of chemical bonds. When these chemical bonds are broken, the energy can be used.

Cell death

Cell death, also called apoptosis, is an essential part of life. As cells become old or broken, they are cleared away and destroyed. Mitochondria help decide which cells are destroyed.

Mitochondria release cytochrome C, which activates caspase, one of the chief enzymes involved in destroying cells during apoptosis.

Because certain diseases, such as cancer, involve a breakdown in normal apoptosis, mitochondria are thought to play a role in the disease.

Storing calcium

Calcium is vital for a number of cellular processes. For instance, releasing calcium back into a cell can initiate the release of a neurotransmitter from a nerve cell or hormones from endocrine cells. Calcium is also necessary for muscle function, fertilization, and blood clotting, among other things.

Because calcium is so critical, the cell regulates it tightly. Mitochondria play a part in this by quickly absorbing calcium ions and holding them until they are needed.

Other roles for calcium in the cell include regulating cellular metabolism, steroid synthesis, and hormone signalingTrusted Source.

Heat production

When we are cold, we shiver to keep warm. But the body can also generate heat in other ways, one of which is by using a tissue called brown fat.

During a process called proton leakTrusted Source, mitochondria can generate heat. This is known as non-shivering thermogenesis. Brown fat is found at its highest levels in babies, when we are more susceptible to cold, and slowly levels reduce as we age.

Mitochondrial disease

The DNA within mitochondria is more susceptible to damage than the rest of the genome.

This is because free radicals, which can cause damage to DNA, are produced during ATP synthesis.

Also, mitochondria lack the same protective mechanisms found in the nucleus of the cell.

However, the majority of mitochondrial diseases are due to mutations in nuclear DNA that affect products that end up in the mitochondria. These mutations can either be inherited or spontaneous.

When mitochondria stop functioning, the cell they are in is starved of energy. So, depending on the type of cell, symptoms can vary widely. As a general rule, cells that need the largest amounts of energy, such as heart muscle cells and nerves, are affected the most by faulty mitochondria.

The following passage comes from the United Mitochondrial Disease Foundation:

> "Because mitochondria perform so many different functions in different tissues, there are literally hundreds of different mitochondrial diseases. [...] Because of the complex

interplay between the hundreds of genes and cells that must cooperate to keep our metabolic machinery running smoothly, it is a hallmark of mitochondrial diseases that identical mtDNA mutations may not produce identical diseases."

Diseases that generate different symptoms but are due to the same mutation are referred to as genocopies.

Conversely, diseases that have the same symptoms but are caused by mutations in different genes are called phenocopies. An example of a phenocopy is Leigh syndrome, which can be caused by several different mutations.

Although symptoms of a mitochondrial disease vary greatly, they might include:

- loss of muscle coordination and weakness
- problems with vision or hearing
- learning disabilities
- heart, liver, or kidney disease
- gastrointestinal problems
- neurological problems, including dementia

Other conditions that are thought to involve some level of mitochondrial dysfunction, include:

- Parkinson's disease
- Alzheimer's disease
- bipolar disorder
- schizophrenia
- chronic fatigue syndrome
- Huntington's disease
- diabetes
- autism

MITOCHONDRIA AND AGING

Over recent years, researchers have investigated a link between mitochondria dysfunction and aging. There are a number of theories surrounding aging, and the mitochondrial free radical theory of aging has become popular over the last decade or so.

The theory is that reactive oxygen species (ROS) are produced in mitochondria, as a byproduct of energy production. These highly charged particles damage DNA, fats, and proteins.

Because of the damage caused by ROS, the functional parts of mitochondria are damaged. When the mitochondria can no longer function so well, more ROS are produced, worsening the damage further.

Although correlations between mitochondrial activity and aging have been found, not all scientists have reached the same conclusions. 'Iheir exact role in the aging process is still unknown.

INTRACELLULAR COMPARTMENTS

The Cell Nucleus

The cell nucleus is one of the largest organelles found in cells and also plays an important biological role. It composes about 10% of the total volume of the cell and is found near the center of eukaryotic cells. Its importance lies in its function as a storage site for DNA, our genetic material. The cell nucleus is composed of two membranes that form a porous nuclear envelope, which allows only select molecules in and out of the cell.

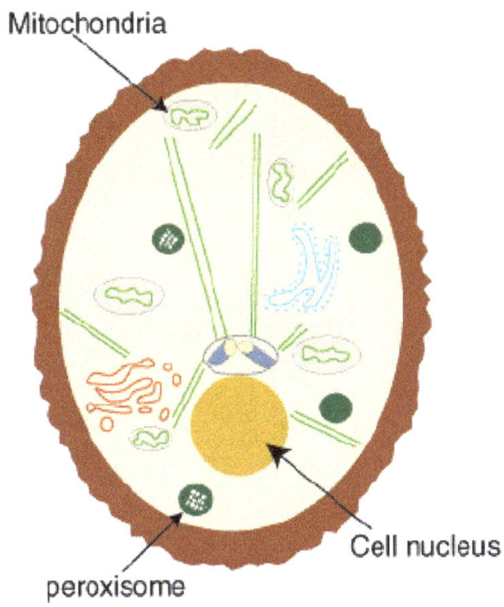

Fig. 2: Location of the cell nucleus, mitochondria, and peroxisomes in a cell.

The DNA that is found in the cell nucleus is packaged into structures called chromosomes. Chromosomes contain DNA and proteins and carry all the genetic information of an organism. The nucleus gains support from intermediate filaments that both form the surrounding nuclear lamina and makes direct contact with the endoplasmic reticulum. The nucleus is also the site of DNA and RNA synthesis.

Mitochondria

The mitochondria, with its specialized double-membrane structure, generate adenosine triphosphate (ATP), a molecule that provides organisms with energy.

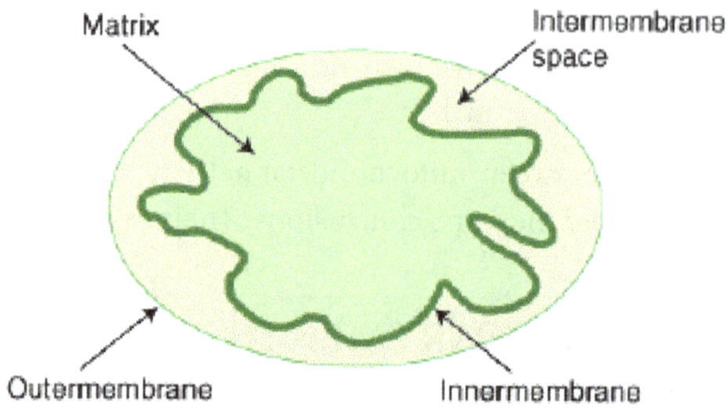

Fig. 3: Mitochondrial structureThe outer and inner membranes of the mitochondria form two sub-compartments: the internal matrix space and the intermembrane space. Those few proteins found withn the mitochondria are located within the inner membrane. Mitochondria synthesize ATP with energy supplied by the electron transport chain and a process called oxidative phosphorylation.

Peroxisomes

Peroxisomes are single-membrane structures found in all eukaryotic cells. They are small, membrane-bound structures that use molecular oxygen to oxidize organic molecules. The structure is one of the major oxygen utilizing organelles, the other being the mitochondria. Peroxisomes contain oxidative enzymes and other enzymes that help produce and degrade hydrogen peroxide.

8

CYTOSKELETON

The cytoskeleton of eukaryotic cells is made of filamentous proteins, and it provides mechanical support to the cell and its cytoplasmic constituents. All cytoskeletons consist of three major classes of elements that differ in size and in protein composition. Microtubules are the largest type of filament, with a diameter of about 25 nanometers (nm), and they are composed of a protein called tubulin. Actin filaments are the smallest type, with a diameter of only about 6 nm, and they are made of a protein called actin. Intermediate filaments, as their name suggests, are mid-sized, with a diameter of about 10 nm. Unlike actin filaments and microtubules, intermediate filaments are constructed from a number of different subunit proteins.

Three major types of filaments make up the cytoskeleton: actin filaments, microtubules, and intermediate filaments. Actin filaments occur in a cell in the form of meshworks or bundles of parallel fibres; they help determine the shape of the cell and also help it adhere to the substrate. The constantly changing arrays of actin filaments help move the cell and mediate specific activities within it, such as cell cleavage during mitosis. Microtubules are longer filaments that are constantly assembling and disassembling; they play a crucial role in moving the daughter chromosomes to the newly forming daughter cells during mitosis, and bundles of microtubules form the cilia and flagella found in protozoans and in the cells of some multicellular animals. Intermediate filaments, in contrast to actin filaments and microtubules, are very stable structures that form the true skeleton of the cell. They anchor the nucleus and position it within the cell, and they give the cell its elastic properties and its ability to withstand tension.

It provides shape and support to the cell, organizes the organelles and facilitates transport of molecules, cell division and cell signalling.

Tubulin contains two polypeptide subunits, and dimers of these subunits string together to make long strands called protofilaments. Thirteen protofilaments then come together to form the hollow, straw-shaped filaments of microtubules. Microtubules are ever-changing, with reactions constantly adding and subtracting tubulin dimers at both ends of the filament. The rates of change at either end are not balanced — one end grows more rapidly and is called the plus end, whereas the other end is known as the minus end. In cells, the minus ends of microtubules are anchored in structures called microtubule organizing centers (MTOCs). The primary MTOC in a cell is called the centrosome, and it is usually located adjacent to the nucleus.

Microtubules tend to grow out from the centrosome to the plasma membrane. In nondividing cells, microtubule networks radiate out from the centrosome to provide the basic organization of the cytoplasm, including the positioning of organelles.

ACTIN FILAMENTS

The protein actin is abundant in all eukaryotic cells. It was first discovered in skeletal muscle, where actin filaments slide along filaments of another protein called myosin to make the cells contract. (In nonmuscle cells, actin filaments are less organized and myosin is much less prominent.) Actin filaments are made up of identical actin proteins arranged in a long spiral chain. Like microtubules, actin filaments have plus and minus ends, with more ATP-powered growth occurring at a filament's plus end.

In many types of cells, networks of actin filaments are found beneath the cell cortex, which is the meshwork of membrane-associated proteins that supports and strengthens the plasma membrane. Such networks allow cells to hold — and move — specialized shapes, such as the brush border of microvilli. Actin filaments are also involved in cytokinesis and cell movement (Fig. 1).

STRUCTURE OF THE CYTOSKELETON

All cells have a cytoskeleton, but usually the cytoskeleton of eukaryotic cells is what is meant when discussing the cytoskeleton. Eukaryotic cells are complex cells that have a nucleus and organelles. Plants, animals, fungi, and protists have eukaryotic cells. Prokaryotic cells are less complex, with no true nucleus or organelles except ribosomes, and they are found in the single-celled organisms bacteria and archaea. The cytoskeleton of prokaryotic cells was originally thought not to exist; it was not discovered until the early 1990s.

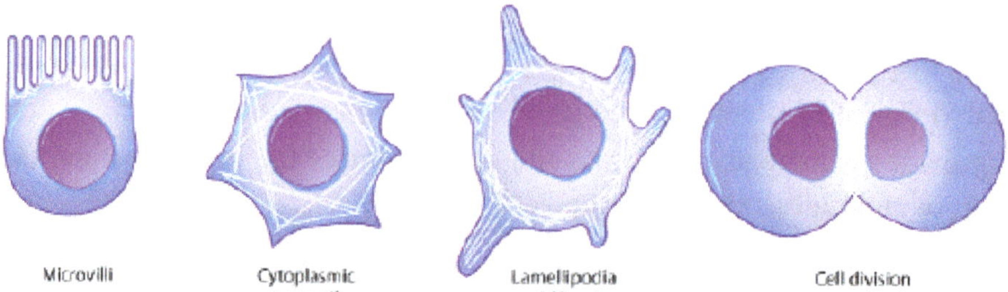

Fig. 1: Actin filaments support a variety of structures in a cell.

The eukaryotic cytoskeleton consists of three types of filaments, which are elongated chains of proteins: microfilaments, intermediate filaments, and microtubules.

The microfilaments of this cell are shown in red, while microtubules are shown in green. The blue dots are nuclei.

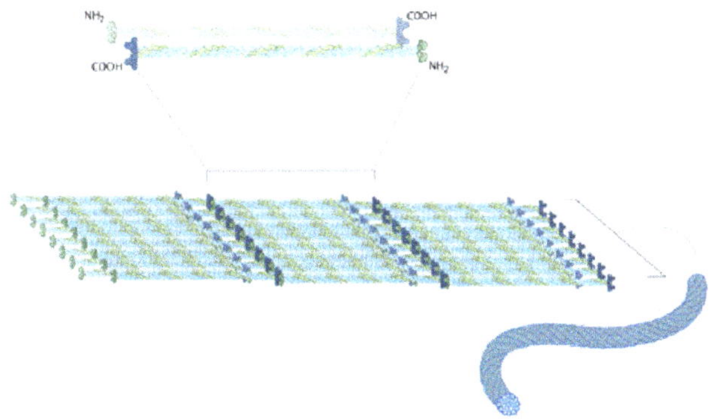

Fig. 2: The structure of intermediate filaments

Microfilaments

Microfilaments are also called actin filaments because they are mostly composed of the protein actin; their structure is two strands of actin wound in a spiral. They are about 7 nanometers thick, making them the thinnest filaments in the cytoskeleton. Microfilaments have many functions. They aid in cytokinesis, which is the division of a cytoplasm of a cell when it is dividing into two daughter cells. They aid in cell motility and allow single-celled organisms like amoebas to move. They are also involved in cytoplasmic streaming, which is the flowing of cytosol (the liquid part of the cytoplasm) throughout the cell. Cytoplasmic streaming transports nutrients and cell organelles. Microfilaments are also part of muscle cells and allow these cells to contract, along with myosin. Actin and myosin are the two main components of muscle contractile elements.

Intermediate Filaments

Intermediate filaments are about 8-12 nm wide; they are called intermediate because they are in-between the size of microfilaments and microtubules. Intermediate filaments are made of different proteins such as keratin (found in hair and nails, and also in animals with scales, horns, or hooves), vimentin, desmin, and lamin. All intermediate filaments are found in the cytoplasm except for lamins, which are found in the nucleus and help support the nuclear envelope that surrounds the nucleus. The intermediate filaments in the cytoplasm maintain the cell's shape, bear tension, and provide structural support to the cell.

All cells have intermediate filaments, but the protein subunits of these structures vary. Some cells have multiple types of intermediate filaments, and some intermediate filaments are associated with specific cell types. For example, neurofilaments are found specifically in neurons (most prominently in the long axons of these cells), desmin filaments are found specifically in muscle cells, and keratins are found specifically in epithelial cells. Other intermediate filaments are distributed more widely. For example, vimentin filaments are found in a broad range of cell types and frequently colocalize with microtubules. Similarly, lamins are found in all cell types, where they form a meshwork that reinforces the inside of the nuclear membrane. Note that intermediate filaments are not polar in the way that actin or tubulin are (Fig. 2).

Microtubules

Microtubules are the largest of the cytoskeleton's fibers at about 23 nm. They are hollow tubes made of alpha and beta tubulin. Microtubules form structures like flagella, which are "tails" that propel a cell forward. They are also found in structures like cilia, which are appendages that increase a cell's surface area and in some cases allow the cell to move. Most of the microtubules in an animal cell come from a cell organelle called the centrosome, which is a microtubule organizing center (MTOC). The centrosome is found near the middle of the cell, and microtubules radiate outward from it. Microtubules are important in forming the spindle apparatus (or mitotic spindle), which separates sister chromatids so that one copy can go to each daughter cell during cell division. They are also involved in transporting molecules within the cell and in the formation of the cell wall in plant cells.

FUNCTION OF THE CYTOSKELETON

As described above, the cytoskeleton has several functions. First, it gives the cell shape. This is especially important in cells without cell walls, such as animal cells, that do not get their shape from a thick outer layer. It can also give the cell movement. The microfilaments and microtubules can disassemble, reassemble, and contract, allowing cells to crawl and migrate, and microtubules help form structures like cilia and flagella that allow for cell movement.

The cytoskeleton organizes the cell and keeps the cell's organelles in place, but it also aids in the movement of organelles throughout the cell. For example, during endocytosis when a cell engulfs a molecule, microfilaments pull the vesicle containing the engulfed particles into the cell. Similarly, the cytoskeleton helps move chromosomes during cell division.

One analogy for the cytoskeleton is the frame of a building. Like a building's frame, the cytoskeleton is the "frame" of the cell, keeping structures in place, providing support, and giving the cell a definite shape.

CELL MOVEMENT

Cytoskeletal filaments provide the basis for cell movement. For instance, cilia and (eukaryotic) flagella move as a result of microtubules sliding along each other. In fact, cross sections of these tail-like cellular extensions show organized arrays of microtubules.

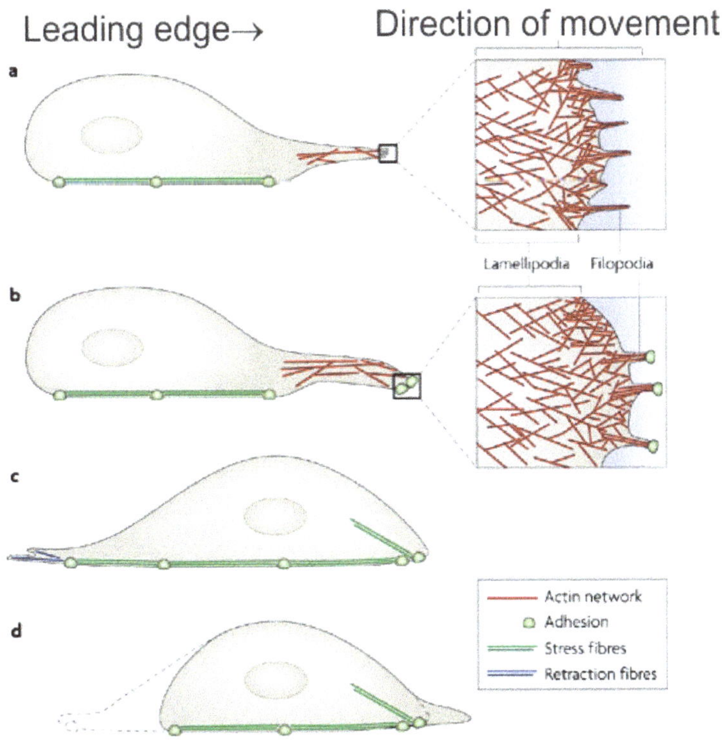

Fig. 3: Cell migration is dependent on different actin filament structures.

(A) In a cell, motility is initiated by an actin-dependent protrusion of the cell's leading edge, which is composed of armlike structures called lamellipodia and filopodia. These protrusive structures contain actin filaments, with elongating barbed ends orientated toward the plasma membrane. (B) During cellular arm extension, the plasma membrane sticks to the surface at the leading edge. (C) Next, the nucleus and the cell body are pushed forward through intracellular contraction forces mediated by stress fibers. (D) Then, retraction fibers pull the rear of the cell forward.

Other cell movements, such as the pinching off of the cell membrane in the final step of cell division (also known as cytokinesis) are produced by the contractile capacity of actin filament networks. Actin filaments are extremely dynamic and can rapidly form and disassemble. In fact, this dynamic action underlies the crawling behavior of cells such as amoebae. At the leading edge of a moving cell, actin filaments are rapidly polymerizing; at its rear edge, they are quickly depolymerizing (Fig. 3). A large number of other proteins participate in actin assembly and disassembly as well.

GENETICS

9

GENETICS: AN OVERVIEW

A gene, the basic unit of heredity, is a segment of DNA containing all the information necessary to synthesize a polypeptide (protein). Protein synthesis, folding, and tertiary and quaternary structure ultimately determine much of the body's structure and function.

HISTORICAL BACKGROUND

Ancient theories of pangenesis and blood in heredity

Although scientific evidence for patterns of genetic inheritance did not appear until Mendel's work, history shows that humankind must have been interested in heredity long before the dawn of civilization. Curiosity must first have been based on human family resemblances, such as similarity in body structure, voice, gait, and gestures. Such notions were instrumental in the establishment of family and royal dynasties. Early nomadic tribes were interested in the qualities of the animals that they herded and domesticated and, undoubtedly, bred selectively. The first human settlements that practiced farming appear to have selected crop plants with favourable qualities. Ancient tomb paintings show racehorse breeding pedigrees containing clear depictions of the inheritance of several distinct physical traits in the horses. Despite this interest, the first recorded speculations on heredity did not exist until the time of the ancient Greeks; some aspects of their ideas are still considered relevant today.

Hippocrates (*c.* 460–*c.* 375 BCE), known as the father of medicine, believed in the inheritance of acquired characteristics, and, to account for this, he devised the hypothesis known as pangenesis. He postulated that all organs of the body of a parent gave off invisible "seeds," which were like miniaturized building components and were transmitted during sexual intercourse, reassembling themselves in the mother's womb to form a baby.

Aristotle (384–322 BCE) emphasized the importance of blood in heredity. He thought that the blood supplied generative material for building all parts of the adult body, and he reasoned that blood was the basis for passing on this generative power to the next generation. In fact, he believed that the male's semen was purified blood and that a woman's menstrual blood was her equivalent of semen. These male and female contributions united in the womb to produce a baby. The blood contained some type of hereditary essences, but he believed that the baby would develop under the influence of these essences, rather than being built from the essences themselves.

Aristotle's ideas about the role of blood in procreation were probably the origin of the still prevalent notion that somehow the blood is involved in heredity. Today people still speak of certain traits as being "in the blood" and of "blood lines" and "blood ties." The Greek model of inheritance, in which a teeming multitude of substances was invoked, differed from that of the Mendelian model. Mendel's idea was that distinct differences between individuals are determined by differences in single yet powerful hereditary factors. These single hereditary factors were identified as genes. Copies of genes are transmitted through sperm and egg and guide the development of the offspring. Genes are also responsible for reproducing the distinct features of both parents that are visible in their children.

TIME LINE OF IMPORTANT MILESTONES IN THE HISTORY OF GENETICS

A time line of important milestones in the history of genetics is provided in the table.

Timeline of important milestones in the history of genetics

Year	Event	
1866	Austrian botanist Gregor Mendel published the results of his experiments with pea plants. His work later provided the mathematical foundation of the science of genetics.	
1869	Swiss biochemist Johann Friedrich Miescher became the first to isolate nuclein-now known as DNA. Although he developed hypotheses explaining the role of nuclein in heredity, he ultimately concluded that one molecule alone could not provide the level of variation observed in nature within and between species.	
1900	Mendel's experiments were rediscovered independently by Dutch botanist and geneticist Hugo de Vries, German botanist and geneticist Carl Erich Correns, and Austrian botanist Erich Tschermak von Seysenegg, giving rise to the modern science of genetics.	
1928	English bacteriologist Frederick Griffith conducted experiments suggesting that bacteria are capable of transferring genetic information and that such transformation is heritable.	
1931	American scientists Harriet B. Creighton and Barbara McClintock published a paper demonstrating that new allelic combinations of linked genes are correlated with physically exchanged chromosome parts. Their findings suggested that chromosomes form the basis of genetics.	
year	**event**	

1944	Canadian-born American bacteriologist Oswald Avery and American biologists Maclyn McCarty and Colin MacLeod reported that the transforming substance-the genetic material of the cell-was DNA.
1950	Austrian-born American biochemist Erwin Chargaff discovered that the components of DNA are paired in a 1:1 ratio. Thus, the amount of adenine (A) is always equal to the amount of thymine (T), and the amount of guanine (G) is always equal to the amount of cytosine (C).
1951	British scientists Rosalind Franklin, Maurice Wilkins, and Raymond Gosling conducted X-ray diffraction studies that provided images of the helical structure of DNA fibres.
1953	Using Chargaff's data and the X-ray images recorded by Franklin, Wilkins, and Gosling, British biophysicists James Watson and Francis Crick determined the molecular structure of DNA. Watson, Crick, and Wilkins shared the 1962 Nobel Prize for Physiology or Medicine for their discovery.
1960s	Swiss microbiologist Werner Arber and American microbiologists Hamilton Othanel Smith and Daniel Nathans discovered restriction enzymes, which cleave DNA into fragments. The discovery, for which the three men shared the 1978 Nobel Prize for Physiology or Medicine, enabled scientists to manipulate genes by removing and inserting DNA sequences.
1970s	American molecular biologists Allan M. Maxam and Walter Gilbert and English biochemist Frederick Sanger developed some of the first techniques for DNA sequencing. Gilbert and Sanger shared the 1980 Nobel Prize for Chemistry for their work.
1983	American biochemist Kary B. Mullis invented the polymerase chain reaction (PCR), a simple technique that allows a specific stretch of DNA to be copied billions of times in a few hours. Mullis received the 1993 Nobel Prize for Chemistry for his invention.
1990	The Human Genome Project (HGP) began. By the time of its completion in 2003, HGP researchers had successfully determined, stored, and rendered publicly available the sequences of almost all the genetic content of the human genome.
2002	The International HapMap Project, which was designed to identify genetic variations contributing to human disease through the development of a haplotype (haploid genotype map of the human genome), began. By completion of Phase II of the project in 2007, scientists had data on some 3.1 million variations in the human genome.
2008	The 1000 Genomes Project, an international collaboration in which researchers aimed to sequence the genomes of a large number of people from different ethnic groups worldwide with the intent of creating a catalog of genetic variations, began. The project was completed in 2015.

PREFORMATION AND NATURAL SELECTION

In the two millennia between the lives of Aristotle and Mendel, few new ideas were recorded on the nature of heredity. In the 17th and 18th centuries the idea of preformation was introduced. Scientists using the newly developed microscopes imagined that they could see miniature replicas of human beings inside sperm heads. French biologist Jean-Baptiste Lamarck invoked the idea of "the inheritance of acquired characters," not as an explanation for heredity but as a model for evolution. He lived at a time when the fixity of species was taken for granted, yet he maintained that this fixity was only found in a constant environment. He enunciated the law of use and disuse, which states that when certain organs become specially developed as a result of some environmental need, then that state of development is hereditary and can be passed on to progeny. He believed that in this way, over many generations, giraffes could arise from deerlike animals that had to keep stretching their necks to reach high leaves on trees.

British naturalist Alfred Russel Wallace originally postulated the theory of evolution by natural selection. However, Charles Darwin's observations during his circumnavigation of the globe aboard the HMS *Beagle* (36–1831) provided evidence for natural selection and his suggestion that humans and animals shared a common ancestry. Many scientists at the time believed in a hereditary mechanism that was a version of the ancient Greek idea of pangenesis, and Darwin's ideas did not appear to fit with the theory of heredity that sprang from the experiments of Mendel.

THE WORK OF MENDEL

Before Gregor Mendel, theories for a hereditary mechanism were based largely on logic and speculation, not on experimentation. In his monastery garden, Mendel carried out a large number of cross-pollination experiments between variants of the garden pea, which he obtained as pure-breeding lines. He crossed peas with yellow seeds to those with green seeds and observed that the progeny seeds (the first generation, F_1) were all yellow. When the F_1 individuals were self-pollinated or crossed among themselves, their progeny (F_2) showed a ratio of 3:1 (3/4 yellow and 1/4 green). He deduced that, since the F_2 generation contained some green individuals, the determinants of greenness must have been present in the F_1 generation, although they were not expressed because yellow is dominant over green.

From the precise mathematical 3:1 ratio (of which he found several other examples), he deduced not only the existence of discrete hereditary units (genes) but also that the units were present in pairs in the pea plant and that the pairs separated during gamete formation. Hence, the two original lines of pea plants were proposed to be *YY* (yellow) and *yy* (green).

The gametes from these were *Y* and *y*, thereby producing an F_1 generation of *Yy* that were yellow in colour because of the dominance of *Y*. In the F_1 generation, half the gametes were *Y* and the other half were *y*, making the F_2 generation produced from random mating 4/1 *Yy*, 1/2 *YY*, and 1/4 *yy*, thus explaining the 3:1 ratio. The forms of the pea colour genes, *Y* and *y*, are called alleles.

Mendel also analyzed pure lines that differed in pairs of characters, such as seed colour (yellow versus green) and seed shape (round versus wrinkled). The cross of yellow round seeds with green wrinkled seeds resulted in an F_1 generation that were all yellow and round, revealing the dominance of the yellow and round traits.

However, the F_2 generation produced by self-pollination of F_1 plants showed a ratio of 16/9) 9:3:3:1 yellow round, 16/3 yellow wrinkled, 16/3 green round, and 16/1 green wrinkled; note that a 9:3:3:1 ratio is simply two 3:1 ratios combined). From this result and others like it, he deduced the independent assortment of separate gene pairs at gamete formation.

endel's success can be attributed in part to his classic experimental approach. He chose his experimental organism well and performed many controlled experiments to collect data. From his results, he developed brilliant explanatory hypotheses and went on to test these hypotheses experimentally. Mendel's methodology established a prototype for genetics that is still used today for gene discovery and understanding the genetic properties of inheritance.

HOW THE GENE IDEA BECAME REALITY

Mendel's genes were only hypothetical entities, factors that could be inferred to exist in order to explain his results. The 20th century saw tremendous strides in the development of the understanding of the nature of genes and how they function. Mendel's publications lay unmentioned in the research literature until 1900, when the same conclusions were reached by several other investigators. Then there followed hundreds of papers showing Mendelian inheritance in a wide array of plants and animals, including humans. It seemed that Mendel's ideas were of general validity. Many biologists noted that the inheritance of genes closely paralleled the inheritance of chromosomes during nuclear divisions, called meiosis, that occur in the cell divisions just prior to gamete formation.

STRUCTURE

Humans have about 20,000 to 23,000 genes depending on how a gene is defined. Genes are contained in chromosomes in the cell nucleus and mitochondria. In humans, somatic (nongerm) cell nuclei normally have 46 chromosomes in 23 pairs. Each pair consists of one chromosome from the mother and one from the father. Twenty-two of the pairs, chromosome numbers 1 to 22, the autosomes, are normally homologous (identical in size, shape, and position and number of genes). The 23rd pair, the sex chromosomes (X and Y), determines a person's sex as well as containing other functional genes. Women have 2 X chromosomes (which are homologous) in somatic cell nuclei; men have one X and one Y chromosome (which are heterologous).

The X chromosome carries genes responsible for many hereditary traits; the smaller Y chromosome carries genes that initiate male sex differentiation, as well as a few other genes. Because the X chromosome has many more genes than the Y chromosome, many X chromosome genes in males are not paired; in order to maintain a balance of genetic material between men and women, one of the X chromosomes in each cell of females is randomly inactivated early in fetal life (lyonization). In some cells, the X from the mother is inactivated, and in others it is the X from the father. Once inactivation has taken place in a cell, all descendants of that cell have the same X inactivation. A karyotype illustrates the full set of chromosomes in a person's cells.

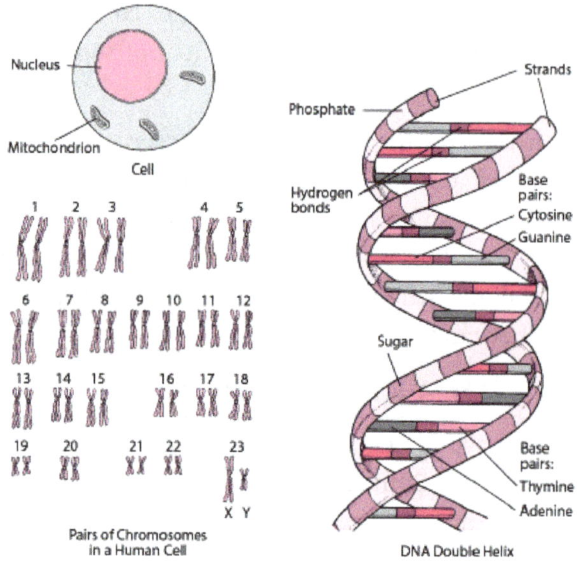

Fig. 1. Structure of DNA

Germ cells (egg and sperm) divide through meiosis, which reduces the number of chromosomes to 23—half the number in somatic cells. In meiosis, the genetic information inherited from a person's mother and father is recombined through crossing over (exchange between homologous chromosomes). When an egg is fertilized by a sperm at conception, the normal number of 46 chromosomes is reconstituted.

Genes are arranged linearly along the DNA of chromosomes. Each gene has a specific location (locus), which is typically the same on each of the 2 homologous chromosomes. The genes that occupy the same locus on each chromosome of a pair (one inherited from the mother and one from the father) are called alleles. Each gene consists of a specific DNA sequence; 2 alleles may have slight differences or the same DNA sequences.

Having a pair of identical alleles for a particular gene is homozygosity; having a pair of nonidentical alleles is heterozygosity. Some genes occur in multiple copies that may be next to each other or in different locations in the same or different chromosomes.

THE DISCOVERY OF LINKED GENES

It seemed that genes were parts of chromosomes. In 1910 this idea was strengthened through the demonstration of parallel inheritance of certain *Drosophila* (a type of fruit fly) genes on sex-determining chromosomes by American zoologist and geneticist Thomas Hunt Morgan. Morgan and one of his students,

Alfred Henry Sturtevant, showed not only that certain genes seemed to be linked on the same chromosome but that the distance between genes on the same chromosome could be calculated by measuring the frequency at which new chromosomal combinations arose (these were proposed to be caused by chromosomal breakage and reunion, also known as crossing over). In 1916 another student of Morgan's, Calvin Bridges, used fruit flies with an extra chromosome to prove beyond reasonable doubt that the only way to explain the abnormal inheritance of certain genes was if they were part of the extra chromosome.

American geneticist Hermann Joseph Müller showed that new alleles (called mutations) could be produced at high frequencies by treating cells with X-rays, the first demonstration of an environmental mutagenic agent (mutations can also arise spontaneously). In 1931 American botanist Harriet Creighton and American scientist Barbara McClintock demonstrated that new allelic combinations of linked genes were correlated with physically exchanged chromosome parts.

GENE FUNCTION

Genes consist of DNA. The length of the gene determines the length of the protein synthesized from the gene code. DNA is a double helix in which nucleotides (bases) are paired:

- Adenine (A) is paired with thymine (T)
- Guanine (G) is paired with cytosine (C)

DNA is transcribed during protein synthesis, in which one strand of DNA is used as a template against which messenger RNA (mRNA) is synthesized. RNA has the same base pairs as DNA, except that uracil (U) replaces thymine (T). Parts of mRNA travel from the nucleus to the cytoplasm and then to a ribosome, where protein synthesis occurs. Transfer RNA (tRNA) brings each amino acid back to the ribosome where it is added to the growing polypeptide chain in a sequence determined by the mRNA.

As a chain of amino acids is assembled, it folds upon itself to create a complex 3-dimensional structure under the influence of nearby chaperone molecules. The code in DNA is written in triplets containing 3 of the 4 possible nucleotides. Specific amino acids are coded by specific triplets termed codons. Because there are 4 nucleotides, the number of possible triplets is 4^3 (64). Because there are only 20 amino acids, there are redundant (extra) triplet combinations. Some triplets code for the same amino acids as other triplets.

Other triplets may code for elements such as instructions to start or stop protein synthesis and the order in which to combine and assemble amino acids. Genes consist of exons and introns. Exons code for amino acid components of the final protein. Introns contain other information that affects control and speed of protein production. Exons and introns together are transcribed onto mRNA, but the segments transcribed from introns are later spliced out. Many factors regulate transcription, including antisense RNA, which is synthesized from the DNA strand that is not transcribed into mRNA. In addition to DNA, chromosomes contain histones and other proteins that affect gene expression (which proteins and how many proteins are synthesized from a given gene).

Genotype refers to a specific genetic composition and sequence; it determines which proteins are coded for production.

Genome refers to the entire composition of a set of haploid chromosomes (single strand), including the genes they contain.

Phenotype refers to the entire physical, biochemical, and physiologic makeup of a person—ie, how the cell (and thus the body) functions. Phenotype is determined by the types and amounts of proteins actually synthesized, ie, how the genes are actually expressed. Specific genotypes may or may not correlate well with phenotype.

Expression refers to the process in which the information encoded in a gene is used to control the assembly of a molecule (usually protein or RNA). Gene expression depends on multiple factors such as whether a trait is dominant or recessive, the penetrance and expressivity of the gene (see Factors Affecting Gene Expression), degree of tissue differentiation (determined by tissue type and age), environmental factors, whether expression is sex-limited or subject to chromosomal inactivation or genomic imprinting, and other unknown factors.

Epigenetic factors

Factors that affect gene expression without changing the genome sequence are epigenetic factors.

Knowledge of the many biochemical mechanisms that mediate gene expression is growing rapidly. One mechanism is variability in intron splicing (also called alternative splicing). Because introns are spliced out, the exons may also be spliced out, and then the exons may be assembled in many combinations, resulting in many different mRNAs capable of coding for similar but different proteins. The number of proteins that can be synthesized by humans is > 100,000 even though the human genome has only about 20,000+ genes.

Other mechanisms mediating gene expression include DNA methylation and histone interactions involving methylation and acetylation. DNA methylation tends to silence gene function. Histone proteins resemble spools around which DNA coils. Histone modifications such as methylation can increase or decrease the quantity of proteins synthesized from a particular gene. Histone acetylation is associated with decreased gene expression. The strand of DNA that is not transcribed to form mRNA may also be used as a template for synthesis of RNA that controls transcription of the opposite strand.

Another important mechanism involves microRNAs (miRNAs). MiRNAs are short, hairpin-derived (hairpin refers to the shape the RNA sequences assume as they bind together) RNAs that repress target gene expression after transcription. They may be involved in regulation of as many as 60% of transcribed proteins.

Traits and Inheritance Patterns

A trait may be as simple as eye color or as complex as susceptibility to diabetes. Expression of a trait may involve one gene or many genes. Some single-gene defects cause abnormalities in multiple tissues, an effect called pleiotropy. For example, osteogenesisimperfecta (a connective tissue disorder that often results from abnormalities in a single collagen gene) may cause fragile bones, deafness, blue-coloredsclerae, dysplastic teeth, hypermobile joints, and heart valve abnormalities.

Construction of a family pedigree

The family pedigree (family tree) is used to diagram inheritance patterns. Pedigrees are commonly used in genetic counseling. The pedigree uses conventional symbols to represent family members and pertinent health information about them (Professional. Fig. # Symbols for constructing a family pedigree). Some familial disorders with identical phenotypes have multiple patterns of inheritance.

GENETIC DISORDERS

Disorders of a Mendelian nature include:

- Haemophilia.
- Sickle Cell Anaemia.
- Phenylketonuria.

Disorders of a chromosomal nature include:

- Down's syndrome.
- Klinefelter's Syndrome.
- Turners Syndrome.

Genetic disorders can happen for many reasons. Genetic disorders often are described in terms of the chromosome that contains the gene that is changed in people who have the disorder.

If the gene is on one of the first 22 pairs of chromosomes, called the autosomes, the genetic disorder is called an autosomal condition. If the gene is on the X chromosome, the disorder is called X-linked.

Genetic disorders also are grouped by how they run in families. Disorders can be dominant or recessive, depending on how they cause conditions and how they run in families.

Dominant

Dominant diseases can be caused by only one copy of a gene having a DNA mutation. If one parent has the disease, each child has a 50% chance of inheriting the mutated gene.

Recessive

For recessive diseases, both copies of a gene must have a DNA mutation in order to get one of these diseases. If both parents have one copy of the mutated gene, each child has a 25% chance of having the disease, even though neither parent has it. In such cases, each parent is called a carrier of the disease. They can pass the disease on to their children, but do not have the disease themselves.

Single Gene Disorders

Some genetic diseases are caused by a DNA mutation in one of a person's genes. For example, suppose part of a gene usually has the sequence TAC. A mutation can change the sequence to TTC in some people. This change in sequence can change the way that the gene works, for example by changing the protein that is made. Mutations can be passed down to a child from his or her parents. Or, they can happen for the first time in the sperm or egg, so that the child will have the mutation but the parents will not. Single gene disorders can be autosomal or X-linked.

For example, sickle cell disease is an autosomal single gene disorder. It is caused by a mutation in a gene found on chromosome 11. Sickle cell disease causes anemia and other complications. Fragile X syndrome, on the other hand, is an X-linked single gene disorder. It is caused by a change in a gene on the X chromosome. It is the most common known cause of intellectual disability and developmental disability that can be inherited (passed from one generation to the next).

CHROMOSOMAL ABNORMALITIES

Different Number of Chromosomes

People usually have 23 pairs of chromosomes. But, sometimes a person is born with a different number. Having an extra chromosome is called trisomy. Missing a chromosome is called monosomy.

For example, people with Down syndrome have an extra copy of chromosome 21. This extra copy changes the body's and brain's normal development and causes intellectual and physical problems for the person. Some disorders are caused by having a different number of sex chromosomes. For example, people with Turner syndromeexternal icon usually have only one sex chromosome, an X. Women with Turner syndrome can have problems with growth and heart defects.

Changes in Chromosomes

Sometimes chromosomes are incomplete or shaped differently than usual. Missing a small part of a chromosome is called a deletion. A translocation is when part of one chromosome has moved to another chromosome. An inversion is when part of a chromosome has been flipped over.

For example, people with Williamssyndromeexternal icon are missing a small part of chromosome 7. This deletion can result in intellectual disability and a distinctive facial appearance and personality.

Complex Conditions

A *complex disease* is caused by both genetic changes and environmental factors. Complex diseases also are called multifactorial. Most chronic diseases, such as heart disease, cancer, and diabetes, are complex conditions. For example, while some cases of cancer are associated with inherited genetic changes, for example, Lynch syndrome and hereditary breast and ovarian cancer, the majority most likely are caused by changes in several genes acting together with environmental exposures.

LAW OF INHERITANCE BY GREGOR MENDEL

Garden Pea (PisumSativum) was the plant that Mendel experimented on for 7 years to get to the point to propose the laws of inheritance in live creatures. Mendel carefully chose seven distinct characteristics of PisumSativum for the investigation concerning hybridization. Mendel used true-breeding lines i.e. those that go through constant self-pollination and display steady characteristic heritance.

PRINCIPLES OF INHERITANCE

When Mendel observed the monohybrid cross he proposed two laws of inheritance-

Law of Dominance –Distinct elements termed as factors control the characteristics. These factors at all times exist as a couple. One of the constituent genes of the couple dominates over the former.

Law of Segregation –Alleles don't blend and the two characteristics are recuperated all through the gamete formation (in the F2 generation). The characters apart from each other and pass on to diverse gametes. Comparable types of gametes are produced by Homozygous and heterozygous produces diverse sorts of a gamete with varied characteristics.

INCOMPLETE DOMINANCE

It is the discovery that was done after Mendel's work. Incomplete dominance is the situation in which both the alleles do not display a dominant trait resulting in a fine combination or a midway amid the characteristics of the alleles.

Codominance

When two alleles lack the dominant-recessive association and thus the duo affects the creature together.

Law of Independent Assortment

Separation of one set of the characteristic is autonomous of the other set of the characters when they are pooled in a hybrid.

THE CHROMOSOMAL THEORY OF INHERITANCE

Both genes and chromosomes exist in sets of two. The homologous chromosome contains the two alleles of a gene pair in the homologous sites. The coupling and split of a set of chromosomes will cause a split in the set of genes (factor) they carry. This united knowledge is termed as the Chromosomal Theory of Inheritance.

SEX DETERMINATION

A particular nuclear arrangement was perceived by Henking. He perceived that this particular nuclear arrangement was found in only fifty per cent of sperms. He termed this body as x. Later it became clear that only Ovum's that obtain only the x chromosome is born female and those that don't have such a case are born male. Thus, the X- chromosome was termed as sex chromosome and the remaining ones were termed as autosomes.

The occurrence due to which a modification in DNA happens and causes a variation in the phenotype and genotype of a creature is termed as a Mutation.

10

MENDELS' LAWS OF INHERITANCE

Inheritance can be defined as the process of how a child receives genetic information from the parent. The whole process of heredity is dependent upon inheritance and it is the reason that the offsprings are similar to the parents. This simply means that due to inheritance, the members of the same family possess similar characteristics.

It was only during the mid 19th century that people started to understand inheritance in a proper way. This understanding of inheritance was made possible by a scientist named Gregor Mendel, who formulated certain laws to understand inheritance known as Mendel's laws of inheritance.

MENDEL'S LAWS OF INHERITANCE

Between 1856-1863, Mendel conducted the hybridization experiments on the garden peas. During that period, he chose some distinct characteristics of the peas and conducted some cross-pollination/ artificial pollination on the pea lines that showed stable trait inheritance and underwent continuous self-pollination. Such pea lines are called true-breeding pea lines.

WHY WAS PEA PLANT SELECTED FOR MENDEL'S EXPERIMENTS?

He selected a pea plant for his experiments:

1. The pea plant can be easily grown and maintained.
2. They are naturally self-pollinating but can also be cross-pollinated.
3. It is an annual plant, therefore, many generations can be studied within a short period of time.
4. It has several contrasting characters.

Mendel conducted 2 main experiments to determine the laws of inheritance. These experiments were:

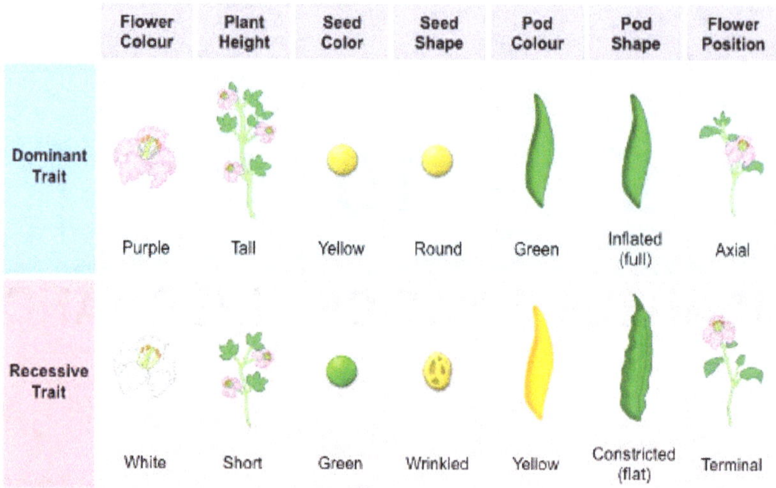

Fig. 1

1. Monohybrid Cross Experiment
2. Dihybrid Cross Experiment

While experimenting, Mendel found that certain factors were always being transferred down to the offspring in a stable way. Those factors are now called genes i.e. genes can be called the units of inheritance.

MENDEL'S EXPERIMENTS

Mendel experimented on a pea plant and considered 7 main contrasting traits in the plants. Then, he conducted both the experiments to determine the aforementioned inheritance laws. A brief explanation of the two experiments is given below.

Monohybrid Cross

In this experiment, Mendel took two pea plants of opposite traits (one short and one tall) and crossed them. He found the first generation offsprings were tall and called it F1 progeny. Then he crossed F1 progeny and obtained both tall and short plants in the ratio 3:1.

Mendel even conducted this experiment with other contrasting traits like green peas vs yellow peas, round vs wrinkled, etc. In all the cases, he found that the results were similar. From this, he formulated the laws of Segregation and Dominance.

Dihybrid Cross

In a dihybrid cross experiment, Mendel considered two traits, each having two alleles. He crossed wrinkled-green seed and round-yellow seeds and observed that all the first generation progeny (F1 progeny) were round-yellow. This meant that dominant traits were the round shape and yellow colour.

He then self-pollinated the F1 progeny and obtained 4 different traits wrinkled-yellow, round-yellow, wrinkled-green seeds and round-green in the ratio 9:3:3:1.

After conducting for other traits, the results were found to be similar. From this experiment, Mendel formulated his second law of inheritance i.e law of Independent Assortment.

Conclusions from Mendel's Experiments

- The genetic makeup of the plant is known as the genotype. On the contrary, the physical appearance of the plant is known as phenotype
- The genes are transferred from parents to the offsprings in pairs known as allele.
- During gametogenesis when the chromosomes are halved, there is a 50% chance of one of the two alleles to fuse with the other parent.
- When the alleles are the same, they are known as homozygous alleles and when the alleles are different they are known as heterozygous alleles.

MENDEL'S LAWS

The two experiments lead to the formulation of Mendel's laws known as laws of inheritance which are:

1. Law of Dominance
2. Law of Segregation
3. Law of Independent Assortment

Law of Dominance

This is also called Mendel's first law of inheritance. According to the law of dominance, hybrid offsprings will only inherit the dominant trait in the phenotype. The alleles that are suppressed are called as the recessive traits while the alleles that determine the trait are known as the dormant traits.

Alleles Can Be Dominant or Recessive

Most familiar animals and some plants have paired chromosomes and are described as diploid. They have two versions of each chromosome: one contributed by the female parent in her ovum and one by the male parent in his sperm. These are joined at fertilization. The ovum and sperm cells (the gametes) have only one copy of each chromosome and are described as haploid.

Mendel's law of dominance states that in a heterozygote, one trait will conceal the presence of another trait for the same characteristic. Rather than both alleles contributing to a phenotype, the dominant allele will be expressed exclusively. The recessive allele will remain "latent," but will be transmitted to offspring by the same manner in which the dominant allele is transmitted. The recessive trait will only be expressed by offspring that have two copies of this allele; these offspring will breed true when self-crossed.

By definition, the terms dominant and recessive refer to the genotypic interaction of alleles in producing the phenotype of the heterozygote. The key concept is genetic: which of the two alleles present in the heterozygote is expressed, such that the organism is phenotypically identical to one of the two homozygotes. It is sometimes convenient to talk about the trait corresponding to the dominant allele as the dominant trait and the trait corresponding to the hidden allele as the recessive trait. However, this can easily lead to confusion in understanding the concept as phenotypic. For example, to say that "green peas" dominate "yellow peas" confuses inherited genotypes and expressed phenotypes. This will subsequently confuse discussion of the molecular basis of the phenotypic difference. Dominance is not inherent. One allele can be dominant to a second allele, recessive to a third allele, and codominant to a fourth. If a genetic trait is recessive, a person needs to inherit two copies of the gene for the trait to be expressed. Thus, both parents have to be carriers of a recessive trait in order for a child to express that trait.

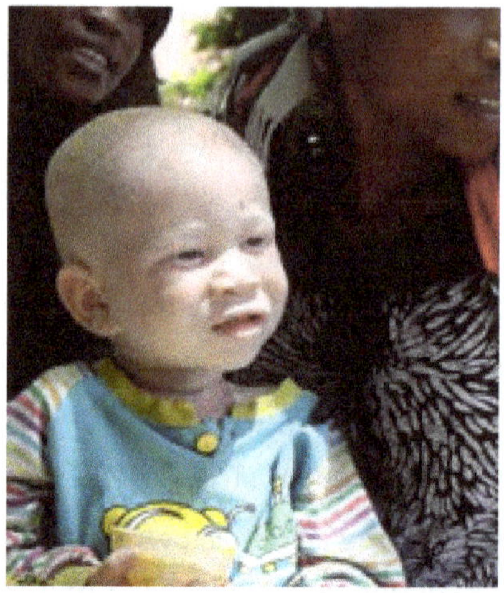

Fig. 2 Recessive traits are only visible if an individual inherits two copies of the recessive allele: The child in the photo expresses albinism, a recessive trait.

Since Mendel's experiments with pea plants, other researchers have found that the law of dominance does not always hold true. Instead, several different patterns of inheritance have been found to exist.

LAW OF SEGREGATION

The law of segregation states that during the production of gametes, two copies of each hereditary factor segregate so that offspring acquire one factor from each parent. In other words, allele (alternative form of the gene) pairs segregate during the formation of gamete and re-unite randomly during fertilization. This is also known as Mendel's third law of inheritance.

Equal Segregation of Alleles

Observing that true-breeding pea plants with contrasting traits gave rise to F_1 generations that all expressed the dominant trait and F_2 generations that expressed the dominant and recessive traits in a 3:1 ratio, Mendel proposed the law of segregation. The law of segregation states that each individual that is a diploid has a pair of alleles (copy) for a particular trait. Each parent passes an allele at random to their offspring resulting in a diploid organism. The allele that contains the dominant trait determines the phenotype of the offspring. In essence, the law states that copies of genes separate or segregate so that each gamete receives only one allele.

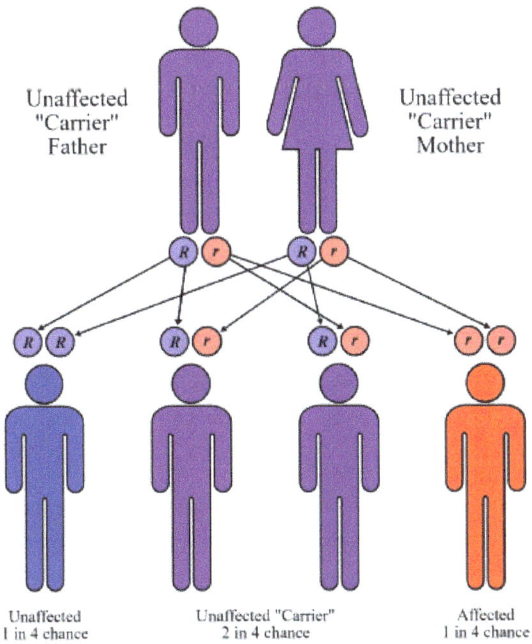

Fig. 3 The Law of Segregation states that alleles segregate randomly into gametes: When gametes are formed, each allele of one parent segregates randomly into the gametes, such that half of the parent's gametes carry each allele.

For the F_2 generation of a monohybrid cross, the following three possible combinations of genotypes could result: homozygous dominant, heterozygous, or homozygous recessive. Because heterozygotes could arise from two different pathways (receiving one dominant and one recessive allele from either parent), and because heterozygotes and homozygous dominant individuals are phenotypically identical, the law supports Mendel's observed 3:1 phenotypic ratio. The equal segregation of alleles is the reason we can apply the Punnett square to accurately predict the offspring of parents with known genotypes.

The physical basis of Mendel's law of segregation is the first division of meiosis in which the homologous chromosomes with their different versions of each gene are segregated into daughter nuclei. The behavior of homologous chromosomes during meiosis can account for the segregation of the alleles at each genetic locus to different gametes. As chromosomes separate into different gametes during meiosis, the two different alleles for a particular gene also segregate so that each gamete acquires one of the two alleles. In Mendel's experiments, the segregation and the independent assortment during meiosis in the F1 generation give rise to the F2 phenotypic ratios observed by Mendel. The role of the meiotic segregation of chromosomes in sexual reproduction was not understood by the scientific community during Mendel's lifetime.

Law of Independent Assortment

Also known as Mendel's second law of inheritance, the law of independent assortment states that a pair of trait segregates independently of another pair during gamete formation. As the individual heredity factors assort independently, different traits get equal opportunity to occur together.

Mendel's law of independent assortment states that genes do not influence each other with regard to the sorting of alleles into gametes: every possible combination of alleles for every gene is equally likely to occur. The independent assortment of genes can be illustrated by the dihybrid cross: a cross between two true-breeding parents that express different traits for two characteristics. Consider the characteristics of seed color and seed texture for two pea plants: one that has green, wrinkled seeds (yyrr) and another that has yellow, round seeds (YYRR). Because each parent is homozygous, the law of segregation indicates that the gametes for the green/wrinkled plant all are yr, while the gametes for the yellow/round plant are all YR. Therefore, the F_1 generation of offspring all are YyRr.

For the F_2 generation, the law of segregation requires that each gamete receive either an R allele or an r allele along with either a Y allele or a y allele. The law of independent assortment states that a gamete into which an r allele sorted would be equally likely to contain either a Y allele or a y allele. Thus, there are four equally likely gametes that can be formed when the YyRr heterozygote is self-crossed as follows: YR, Yr, yR, and yr. Arranging these gametes along the top and left of a 4 × 4 Punnett square gives us 16 equally likely genotypic combinations. From these genotypes, we infer a phenotypic ratio of 9 round/yellow:3 round/green:3 wrinkled/yellow:1 wrinkled/green. These are the offspring ratios we would expect, assuming we performed the crosses with a large enough sample size.

Because of independent assortment and dominance, the 9:3:3:1 dihybrid phenotypic ratio can be collapsed into two 3:1 ratios, characteristic of any monohybrid cross that follows a dominant and recessive pattern. Ignoring seed color and considering only seed texture in the above dihybrid cross, we would expect that three-quarters of the F_2 generation offspring would be round and one-quarter would be wrinkled. Similarly, isolating only seed color, we would assume that three-quarters of the F_2 offspring would be yellow and one-quarter would be green. The sorting of alleles for texture and color are independent events, so we can apply the product rule. Therefore, the proportion of round and yellow F_2 offspring is expected to be (16/9 = (4/3) × (4/3, and the proportion of wrinkled and green offspring is expected to be (16/1 = (4/1) × (4/1. These proportions are identical to those obtained using a Punnett square. Round/green and wrinkled/yellow offspring can also be calculated using the product rule as each of these genotypes includes one dominant and one recessive phenotype. Therefore, the proportion of each is calculated as (16/3 = (4/1) × (4/3.

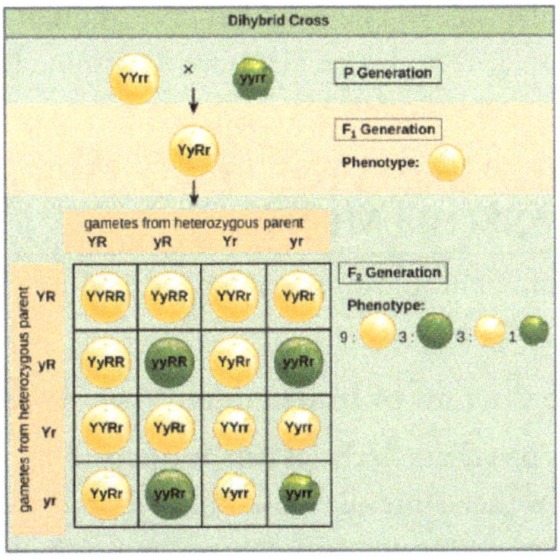

Fig. 4 Independent assortment of 2 genes: This dihybrid cross of pea plants involves the genes for seed color and texture.

FORKED-LINE METHOD

When more than two genes are being considered, the Punnett-square method becomes unwieldy. For instance, examining a cross involving four genes would require a 16 × 16 grid containing 256 boxes. It would be extremely cumbersome to manually enter each genotype. For more complex crosses, the forked-line and probability methods are preferred.

To prepare a forked-line diagram for a cross between F_1 heterozygotes resulting from a cross between AABBCC and aabbcc parents, we first create rows equal to the number of genes being considered and then segregate the alleles in each row on forked lines according to the probabilities for individual monohybrid crosses. We then multiply the values along each forked path to obtain the F_2 offspring probabilities. Note that this process is a diagrammatic version of the product rule. The values along each forked pathway can be multiplied because each gene assorts independently. For a trihybrid cross, the F_2 phenotypic ratio is 27:9:9:9:3:3:3:1.

PROBABILITY METHOD

While the forked-line method is a diagrammatic approach to keeping track of probabilities in a cross, the probability method gives the proportions of offspring expected to exhibit each phenotype (or genotype) without the added visual assistance.

To fully demonstrate the power of the probability method, however, we can consider specific genetic calculations. For instance, for a tetrahybrid cross between individuals that are heterozygotes for all four genes, and in which all four genes are sorting independently in a dominant and recessive pattern, what proportion of the offspring will be expected to be homozygous recessive for all four alleles? Rather than writing out every possible genotype, we can use the probability method. We know that for each gene the fraction of homozygous recessive offspring will be 1/4. Therefore, multiplying this fraction for each of the four genes, (1/4) × (1/4) × (1/4) × (1/4), we determine that 1/256 of the offspring will be quadruply homozygous recessive.

GENETIC LINKAGE AND VIOLATION OF THE LAW OF INDEPENDENT ASSORTMENT

Genes that are on the same chromosome, or "linked", do not assort independently, but can be separated by recombination.

Linked Genes Violate the Law of Independent Assortment

Although all of Mendel's pea characteristics behaved according to the law of independent assortment, we now know that some allele combinations are not inherited independently of each other. Genes that are located on separate non-homologous chromosomes will always sort independently. However, each chromosome contains hundreds or thousands of genes organized linearly on chromosomes like beads on a string. The segregation of alleles into gametes can be influenced by linkage, in which genes that are located physically close to each other on the same chromosome are more likely to be inherited as a pair. However, because of the process of recombination, or "crossover," it is possible for two genes on the same chromosome to behave independently, or as if they are not linked. To understand this, let's consider the biological basis of gene linkage and recombination.

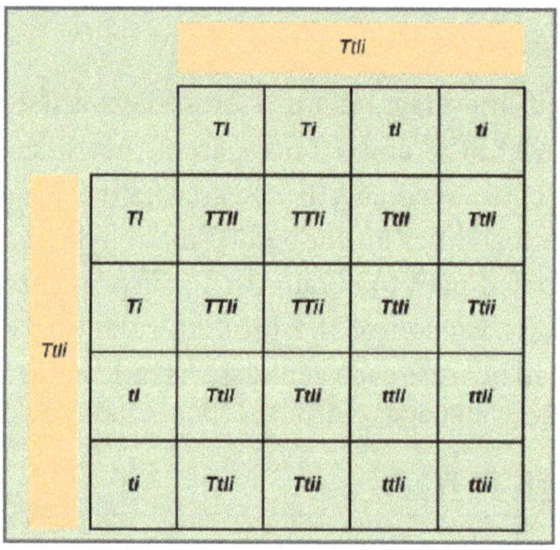

Fig. 5 Unlinked genes assort independently: This figure shows all possible combinations of offspring resulting from a dihybrid cross of pea plants that are heterozygous for the tall/dwarf and inflated/constricted alleles.

Homologous chromosomes possess the same genes in the same linear order. The alleles may differ on homologous chromosome pairs, but the genes to which they correspond do not. In preparation for the first division of meiosis, homologous chromosomes replicate and synapse. Like genes on the homologs align with each other. At this stage, segments of homologous chromosomes exchange linear segments of genetic material. This process is called recombination, or crossover, and it is a common genetic process. Because the genes are aligned during recombination, the gene order is not altered. Instead, the result of recombination is that maternal and paternal alleles are combined onto the same chromosome. Across a given chromosome, several recombination events may occur, causing extensive shuffling of alleles.

When two genes are located in close proximity on the same chromosome, they are considered linked, and their alleles tend to be transmitted through meiosis together. To exemplify this, imagine a dihybrid cross involving flower color and plant height in which the genes are next to each other on the chromosome. If one homologous chromosome has alleles for tall plants and red flowers, and the other chromosome has genes for short plants and yellow flowers, then when the gametes are formed, the tall and red alleles will go together into a gamete and the short and yellow alleles will go into other gametes. These are called the parental genotypes because they have been inherited intact from the parents of the individual producing gametes.

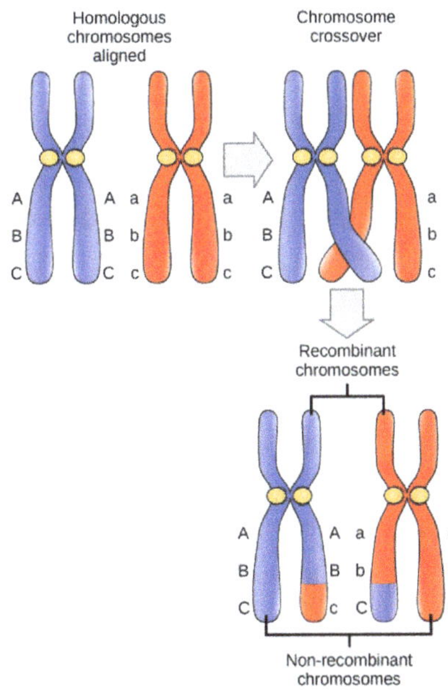

Fig. 6 Linked genes can be separated by recombination: The process of crossover, or recombination, occurs when two homologous chromosomes align during meiosis and exchange a segment of genetic material. Here, the alleles for gene C were exchanged. The result is two recombinant and two non-recombinant chromosomes.

But unlike if the genes were on different chromosomes, there will be no gametes with tall and yellow alleles and no gametes with short and red alleles. If you create the Punnett square with these gametes, you will see that the classical Mendelian prediction of a 9:3:3:1 outcome of a dihybrid cross would not apply. As the distance between two genes increases, the probability of one or more crossovers between them increases, and the genes behave more like they are on separate chromosomes. Geneticists have used the proportion of recombinant gametes (the ones not like the parents) as a measure of how far apart genes are on a chromosome.

Using this information, they have constructed elaborate maps of genes on chromosomes for well-studied organisms, including humans. Mendel's seminal publication makes no mention of linkage, and many researchers have questioned whether he encountered linkage, but chose not to publish those crosses out of concern that they would invalidate his independent assortment postulate. The garden pea has seven chromosomes and some have suggested that his choice of seven characteristics was not a coincidence.

However, even if the genes he examined were not located on separate chromosomes, it is possible that he simply did not observe linkage because of the extensive shuffling effects of recombination.

EPISTASIS

Epistasis occurs when one gene masks or interferes with the expression of another.

Mendel's studies in pea plants implied that the sum of an individual's phenotype was controlled by genes (or as he called them, unit factors): every characteristic was distinctly and completely controlled by a single gene. In fact, single observable characteristics are almost always under the influence of multiple genes (each with two or more alleles) acting in unison. For example, at least eight genes contribute to eye color in humans. In some cases, several genes can contribute to aspects of a common phenotype without their gene products ever directly interacting.

In the case of organ development, for instance, genes may be expressed sequentially, with each gene adding to the complexity and specificity of the organ. Genes may function in complementary or synergistic fashions: two or more genes need to be expressed simultaneously to affect a phenotype. Genes may also oppose each other with one gene modifying the expression of another. In epistasis, the interaction between genes is antagonistic: one gene masks or interferes with the expression of another. "Epistasis" is a word composed of Greek roots that mean "standing upon."

The alleles that are being masked or silenced are said to be hypostatic to the epistatic alleles that are doing the masking. Often the biochemical basis of epistasis is a gene pathway in which the expression of one gene is dependent on the function of a gene that precedes or follows it in the pathway.

An example of epistasis is pigmentation in mice. The wild-type coat color, agouti (AA), is dominant to solid-colored fur (aa). However, a separate gene (C) is necessary for pigment production. A mouse with a recessive c allele at this locus is unable to produce pigment and is albino regardless of the allele present at locus A. Therefore, the genotypes AAcc, Aacc, and aacc all produce the same albino phenotype. A cross between heterozygotes for both genes (AaCc x AaCc) would generate offspring with a phenotypic ratio of 9 agouti:3 solid color:4 albino. In this case, the C gene is epistatic to the A gene.

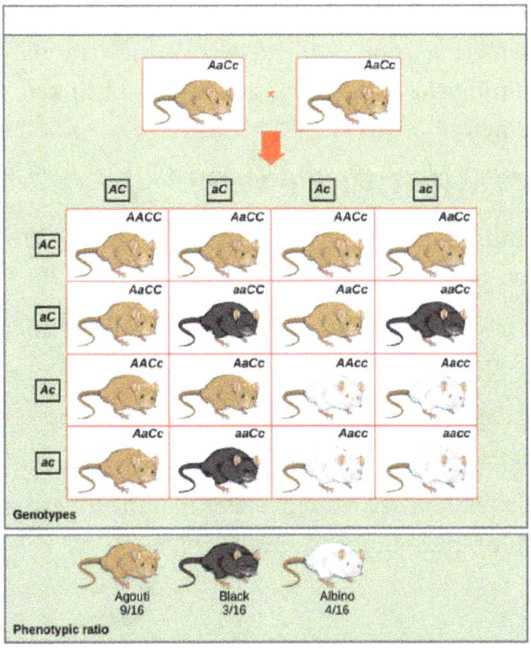

Fig. 7 Epistasis in mouse coat color: In mice, the mottled agouti coat color (A) is dominant to a solid coloration, such as black or gray. A gene at a separate locus (C) is responsible for pigment production. The recessive c allele does not produce pigmentnand a mouse with the homozygous recessive cc genotype is albino regardless of the allele present at the A locus. Thus, the C gene is epistatic to the A gene.

Epistasis can also occur when a dominant allele masks expression at a separate gene. Fruit color in summer squash is expressed in this way. Homozygous recessive expression of the W gene (ww) coupled with homozygous dominant or heterozygous expression of the Y gene (YY or Yy) generates yellow fruit, while the wwyy genotype produces green fruit. However, if a dominant copy of the W gene is present in the homozygous or heterozygous form, the summer squash will produce white fruit regardless of the Y alleles. A cross between white heterozygotes for both genes (WwYy × WwYy) would produce offspring with a phenotypic ratio of 12 white:3 yellow:1 green.

Finally, epistasis can be reciprocal: either gene, when present in the dominant (or recessive) form, expresses the same phenotype. In the shepherd's purse plant (*Capsella bursa-pastoris*), the characteristic of seed shape is controlled by two genes in a dominant epistatic relationship. When the genes A and B are both homozygous recessive (aabb), the seeds are ovoid. If the dominant allele for either of these genes is present, the result is triangular seeds. That is, every possible genotype other than aabb results in triangular seeds; a cross between heterozygotes for both genes (AaBb x AaBb) would yield offspring with a phenotypic ratio of 15 triangular:1 ovoid.

Keep in mind that any single characteristic that results in a phenotypic ratio that totals 16 is typical of a two-gene interaction. Recall the phenotypic inheritance pattern for

Mendel's dihybrid cross, which considered two non-interacting genes: 9:3:3:1. Similarly, we would expect interacting gene pairs to also exhibit ratios expressed as 16 parts. Note that we are assuming the interacting genes are not linked; they are still assorting independently into gametes.

KEY POINTS ON MENDEL'S LAWS

- The law of inheritance was proposed by Gregor Mendel after conducting experiments on pea plants for seven years.
- The Mendel's laws of inheritance include law of dominance, law of segregation and law of independent assortment.
- The law of segregation states that every individual possesses two alleles and only one allele is passed on to the offspring.
- The law of independent assortment states that the inheritance of one pair of genes is independent of inheritance of another pair.

11

LETHALITY AND INTERACTION OF GENES

Lethals or lethal genes or lethal alleles are alleles causing the end of an entity which carries it. Basically, lethal genes are lethal to the organism carrying it, lethal meaning (here) death. Typically it is the consequence of gene-mutation that is required to grow and develop. In 1905, French geneticist Lucien Cuénot was the first person to discover Lethal genes when he was carrying out research on inheritance of the coat color in mice. As per Cuenot, color yellow was dominant over brown presided over by the single gene "Y". Hence, he found that under homozygous conditions, mice could never be yellow in color.

In 1905, Lucien Cuénot observed unusual patterns when studying inheritance of a coat color gene in mice. After mating two yellow mice, he observed that the offspring never showed a normal 3:1 phenotypic ratio. Instead, Cuénot always observed a 2:1 ratio, with two yellow mice for every one non-yellow mouse. Cuénot thus determined that yellow coat color was the dominant phenotypic trait, and by using test crosses, he showed that all his yellow mice were heterozygotes.

Shortly thereafter, in 1910, W. E. Castle and C. C. Little confirmed Cuénot's unusual segregation ratios (Figure 1). Moreover, they demonstrated that Cuénot's crosses resulted in what appeared to be non-Mendelian ratios because he had discovered a lethal gene. Castle and Little did this by showing that one-quarter of the offspring from crosses between heterozygotes died during embryonic development. This was why Cuénot never observed homozygous yellow mice! Thus, by considering embryonic lethality, or death, as a new phenotypic class, the classic 1:2:1 Mendelian ratio of genotypes could be reestablished.

As these examples illustrate, lethal genes cause the death of the organisms that carry them. Sometimes, death is not immediate; it may even take years, depending on the gene. In any case, if a mutation results in lethality, then this is indicative that the affected gene has a fundamental function in the growth, development, and survival of an organism.

The nature of lethal genes can be dominant, recessive or even conditional, depending on the participating genes. Death of an entity can be caused at any time, the most common occurrence is in the early stages of development.

There are genes which control certain phenotypic traits, and at the same time also influence the viability of the individuals. For jnstance, white eyed and vestigial winged flies have lower viabilities than the wild types. There are still other genes which have no effect on the phenotype but influence the viability. This influence on viability may be of such an order that the individual may fail to survive. Such genes which cause the death of the individual carrying it are known as lethal genes.

If the lethal effect is dominant over its normal allele, all individuals carrying it will die and the gene cannot be transmitted to next generation. Such dominant lethal genes will, therefore, be lost in the same generation. On the other hand, recessive lethals are carried in heterozygous condition, and will express themselves only in the homozygous condition. In *Drosophila* genes like curly wings *(Cy)*, plum eyes *(Pm)* and stubble bristles *(Sb)* influence the viability of the flies when present in homozygous condition.

For instance, an individual *Cycy* (heterozygous) will have curly wings and will survive. However, homozygous dominant i.e., *CyCy* will not survive. These genes, therefore, have dual effect. They are dominant with respect to the phenotype (e.g. curly wings, plum eyes and stubble bristles) but are recessive with respect to the lethal effect (consult Mutations : 1. Morphological Level (Including Lethal Mutations)). In 1905, a French geneticist, L. Cuenot reported on the inheritance of mouse body colour, which did not fit the expected Mendelian segregation pattern. It was shown that yellow body colour was dominant over normal brown colour and was controlled by a single gene which may be designated as *Y*.

It was fowid that yellow mice could never be obtained in homozygous condition. When yellow mice were crossed among themselves, segregation for yellow and brown body colour was obtained in 2 : 1 ratio. The brown individuals were pure and were, therefore, homozygous and yellow individuals as usual were heterozygous. These results could be explained on the assumption that the dominant allele for yellow body colour is lethal in homozygous condition. In other words, we may say that, while the gene *Y* is dominant for its effect on the body colour, it is recessive for lethality. Consequently whenever a homozygous individual for *Y* is produced, the lethal character will express itself and the individual will die. Homozygous yellow will thus never be produced. If *Y* is the dominant allele for yellow body colour and *y* is its recessive allele, the results can be expressed as in Fig. 1.

Soon after Cuenot's discovery in mouse, E. Baur in 1907 reported a lethal gene in snapdragon, *Antirrhinum majus.* He found that when yellow leaved dominant genotype called aurea was crossed with its own type (aurea x aurea), normal green plants and aurea heterozygotes were obtained. Aurea homozygotes were though obtained, but lacked the ability to make chlorophyll so that they died before germination or as seedlings.

A large number of such lethal factors are known in a variety of individuals. For instance in chickens, the character 'creeper' is dominant for its effect on phenotype but is recessive for lethality.

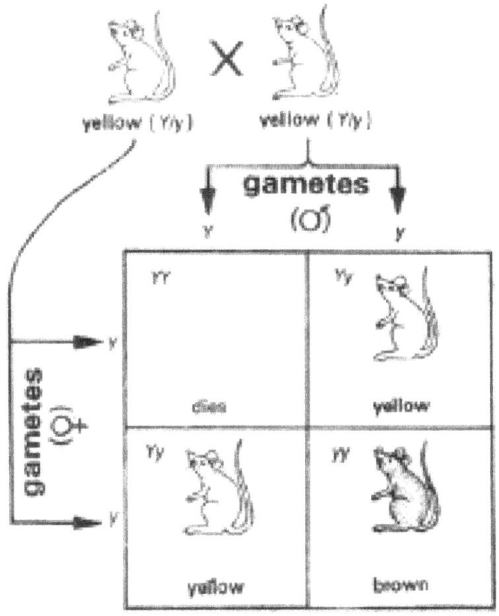

In human beings also there are diseases like infantile amaurotic idiocy and thallassemia major which are associated with lethal factors. Similarly in large number of plant materials, recessive Iethals have been reported. These lethals in plants mostly exhibit themselves through loss of capacity to produce chlorophyll.

PENETRANCE OF LETHALS

The genes causing lethality may differ in the level of their penetrance and expressivity. This results in a broad range of effects ranging from a condition where no survival is observed among affected genotypes, to those where almost complete viability is observed. Many of these genes which do not cause definite lethality are called semilethals or subvitals. It is difficult to draw a line between categories of lethals, but usually only those genes are regarded as lethal genes, which cause the death of the organism at an early stage.

BALANCED LETHAL SYSTEMS

Itwas shown fairly early, that lethal genes can be used to develop a balanced lethal system, where both homozygotes, dominant and recessive will c e so that the surviving phenotype will always be the same and will consist of heterozygotes only. Such balanced lethal systems are known in *Drosophila,* mice and *Oenothera.*

TYPES OF LETHAL GENES

Classification of lethal genes are as follows:

Recessive lethals

These are commonly found, as most lethal are recessive. Their expression is in homozygous conditions only, hence survival of heterozygous ones is unaffected. They are a pair of identical alleles that ultimately result in the death of an entity. Despite the fact that recessive lethals can code for recessive or dominant characteristics, they turn fatal only under homozygous conditions. Example of recessive lethals is as seen in the case of a Manx cat.

In 1907, Edwin Baur began his work with the snapdragon plant *Antirrhinum* and characterized the condition *aurea*, in which plants had golden instead of green leaves. When two *aurea* snapdragon plants were crossed, Baur observed a 2:1 ratio of green seedlings to yellow seedlings. Homozygous *aurea* plants lacked normal chlorophyll development and died either during the embryonic stage or when the plant seedlings were two to three days old. In other words, like Cuénot's homozygous mice, the homozygous *aurea* plants could not fully develop, so an entire class of progeny died.

Cuénot and Baur discovered these first recessive lethal genes because they altered Mendelian inheritance ratios. Recessive lethal genes can code for either dominant or recessive traits, but they do not actually cause death unless an organism carries two copies of the lethal allele. Examples of human diseases caused by recessive lethal alleles include cystic fibrosis, sickle-cell anemia, and achondroplasia. Achondroplasia is an autosomal dominant bone disorder that causes dwarfism. While the inheritance of one achondroplasia allele can cause the disease, the inheritance of two recessive lethal alleles is fatal.

Dominant lethals

These are the alleles whose presence is required in one copy in an entity for them to turn fatal. These are not so frequently found as they cause the death of an entity before they are transmitted to their offspring. A good example of dominant lethal alleles that is seen in humans is the rare - Huntington's disease. It is a neurodegenerative disorder which results in death ultimately. One more example in humans is epiloia genes.

Conditional lethals

These alleles turn deadly only when there is an external environmental aspect involved. For instance, Favism is a conditional lethal gene. This sex-inherited condition causes the carrier to develop hemolyticanemia upon consuming fava beans.Favism is a sex-linked, inherited condition that results from deficiency in an enzyme called glucose-6-phosphate dehydrogenase. It is most common among people of Mediterranean, African, Southeast Asian, and Sephardic Jewish descent. The disease was named because when affected individuals eat fava beans, they develop hemolyticanemia, a condition in which red blood cells break apart and block blood vessels. Blockage can cause kidney failure and result in death. Affected individuals may also develop anemia when administered therapeutic doses of antimalarial medications and other drugs. Note, however, that the defective glucose-6-phosphate dehydrogenase allele only causes death under certain conditions, which makes it a conditional lethal gene. But why would this allele be so common? The interesting thing about individuals with the favism allele is that they are resistant to malaria, because it is more difficult for malaria parasites to multiply in cells with deficient amounts of glucose-6-phosphate dehydrogenase. Therefore, carrying the allele for favism confers an intrinsic genetic or adaptive advantage by protecting individuals from contracting malaria.

Conditional lethal genes can also be expressed due to specific circumstances, such as temperature. For example, a mutant protein may be genetically engineered to be fully functional at 30°C and completely inactive at 37°C. Meanwhile, the wild-type protein is fully functional at both temperatures. The condition in which the mutant phenotype is expressed is termed nonpermissive. Meanwhile, the condition in which the wild-type phenotype is expressed is called permissive. In order to study a conditional lethal mutant, the organism must be maintained under permissive conditions and then switched to the nonpermissive condition during the course of a specific experiment. By developing a conditional lethal version of a dominant lethal gene, scientists can study and maintain organisms carrying dominant lethal alleles.

Balanced lethals

In a self permanent stock, the balancing effect between two different lethals is the balanced lethal system. The lethal genes that are associated in the repulsion stage of linkage are balanced lethals. They sustain themselves in the repulsion phase as a result of tight linkage. The recessive alleles in the repulsive stage of one gene and the dominant allele of the other gene are found to be present in the same chromosome. This lethal system sustains genes associated closely with the lethal genes in a permanent heterozygous state. Such lethals are observed in Drosophila, mice, etc.

Gameticlethals

Gameticlethals are the ones that make the gametes incompetent to fertilize. Meiotic drive is often used to describe these types of lethals. This drive is a series of events which causes the production of unequal numbers by a heterozygote of functional gametes.

CHEMICAL SYNTHETIC LETHALITY

Chemical synthetic lethality/sickness screening encompasses the combination of genetic perturbations with chemicals, as well combinations of chemicals with each other, and observing the pairs that lead to a more extreme growth defect than expected based on the individual effects of the perturbations. Although genetic synthetic lethality can be approached from a solely biological discovery perspective, as soon as chemicals are introduced into the equation, the applications become increasingly about drug discovery and combination therapies, as well as probing drug mechanism of action. We will attempt to discuss the two elements of chemical synthetic lethality, chemical-genetic interactions and chemical-chemical interactions, separately; however, many times these approaches are used in conjunction with each other, as well as tied back to genetic synthetic lethality. Much of the methodology from the genetic synthetic lethality section above can be adapted for use in chemical-genetic synthetic lethality studies.

Chemical and chemical-genetic approaches rely on the use of chemicals to probe bacterial gene functions. In addition to genetic approaches to study essential genes such as knockdowns and conditional essentiality, chemicals are another way to probe these genes. The amount of a chemical probe introduced can be both varied and reversed. Chemical-genetic synthetic lethal/sick interactions can be studied by applying the approaches used for genetic synthetic lethality studies, and simply replacing one of the genetic perturbations with a chemical perturbation.

Furthermore, chemical–chemical synthetic lethality/sickness can be probed in the same way but using only chemical perturbations, without genetic disruption. Often times the chemicals used in chemical–chemical probing may be selected based on the discoveries of chemical–genetic or genetic synthetic lethality/sickness. In this section, we summarize the main approaches that have been used to explore chemical synthetic lethal/sick interactions in bacteria.

Chemical-genetic synthetic lethality

The concept of synthetic lethal/sick chemical screening has long been used in the field of cancer research to screen for cancer drugs. The premise has also been applied to bacterial systems. One of the earlier chemical synthetic lethality screens performed in bacteria was done to probe the mechanism by which the loss of 2-C-methyl-D-erythritol-2,4-cyclodiphosphate (MEC) synthase results in cell death. Campbell and Brown screened MEC synthase-depleted *E. coli* and *B. subtilis* cells against a panel of antibiotics and observed which antibiotics had a lowered minimum inhibitory concentration (MIC) in MEC synthase depletion. Synthetic lethal/sick interactions were observed between MEC synthase and cell wall inhibitors including ampicillin, bacitracin, cycloserine, fosfomycin and vancomycin, as well as with fosmidomycin, an inhibitor of the 1-deoxy-D-xylulose 5-phosphate pathway). Since then, chemical synthetic lethality screening has gotten exponentially higher throughput.

Transposon sequencing-based approaches

The use of TIS and other transposon-based techniques in combination with chemicals to probe synthetic lethality has been of utmost importance of the field of chemical genomics. These approaches can be applied to chemical synthetic lethality studies in a similar manner to how they are used to probe genetic synthetic lethality. Rather than performing the transposon mutagenesis in a strain with a query gene deletion, it is performed both in the presence and absence of exposure to a chemical.

Hypersensitivity of *Mycobacterium smegmatis* and *M. tuberculosis* to β–lactams was characterized using transposon mutagenesis followed by sequencing. Under normal circumstances mycobacteria are resistant to β-lactams; however, this methodology allowed the exploration of instances in which gene disruptions resulted in sensitivity to ceftriaxone, a β–lactam, in β–lactamase mutants. Six mutants with a single transposon insertion were identified in *M. smegmatis* and three in *M. tuberculosis*.

Two of these gene disruptions resulting in synthetic lethality/sickness with ceftriaxone were involved in peptidoglycan biosynthesis—*ponA2* and *dapB*—others were involved in cell division and other cell envelope related processes. Likewise, synthetic lethality of transposon gene disruptions with the aminoglycoside antibiotic tobramycin was also studied by TIS experiments. Screening for differences in the transposon read counts throughout the genome of the untreated and tobramycin-treated transposon mutants led to the identification of 117 genes that were synthetic lethal/sick with sub-inhibitory levels of tobramycin.

Mutants were tested downstream for changes in the MIC of tobramycin and those with strong hypersensitivity to tobramycin included genes involved in the folding of membrane proteins and cell envelope stress proteins, likely causing the cells to no longer tolerate tobramycin-induced aberrantly folded proteins. Another study examining synthetic lethality/sickness to explore antibiotic resistance determinants was performed in two different strains of *Streptococcus pneumoniae*.

The group performed TIS on pathogenic strain TIGR4, both untreated and treated with daptomycin. They discovered that mutations in peptidoglycan biosynthetic or regulatory genes and genes encoding structural membrane proteins were synthetic lethal/sick with daptomycin, suggesting that daptomycin destabilizes the envelope. A potassium ion uptake system was also found to be synthetic lethal/sick with daptomycin, indicating that daptomycin induces potassium ion leaking via membrane disruption. However, many of the interactions were different in another strain of *S. pneumoniae*, Taiwan-19F; therefore, synthetic lethality/sickness can be strain specific. TIS can thus provide the opportunity to increase our understanding of known and unknown drug mechanisms of action and whether they have different impacts on different strains.

Identification of synthetic lethal/sick interactions of *S. aureus* genes with multiple antibiotics has also been explored. TIS was performed to assess mutant fitness when treated with ciprofloxacin, daptomycin, gentamicin, linezolid, oxacillin and vancomycin, relative to the untreated control. Mutations in *sigB* and *pbp4* resulted in sensitization to oxacillin, while mutations in *vraTSR* resulted in sensitivity to vancomycin; these genes are all involved in cell wall integrity. In fact, 21 genes were discovered to be synthetic lethal/sick with at least two of the antibiotics tested, including two uncharacterized genes: *SAOUHSC_01025* and *SAOUHSC_01050*, which were hits for vancomycin, daptomycin and oxacillin. Their antibiotic sensitivity profiles also most closely matched those of *graS*, *mprF* and *cvfC*, indicating a role for these genes in envelope integrity.

Chemical–genetic synthetic lethality has also been studied in *S. aureus* to increase understanding of wall teichoic acid (WTA) function. Walker's group performed a TIS synthetic lethality screen with and without exposure to the TarO inhibitor tunicamycin, resulting in a knockdown of WTA synthesis. A majority of the mutants found to be synthetic lethal/sick with tunicamycin have gene products known to localize to or be involved in the cell envelope. This is unsurprising since WTA is a component of the cell envelope. Furthermore, genes encoding the GraRS-VraFG complex were synthetic lethal/sick with TarO inhibition by tunicamycin. This complex regulates expression of *dltABCD*, known to D-alanylate WTAs and lipoteichoic acids (LTAs), indicating that D-alanylation of LTAs is required in WTA-deficient strains.

Other bacterial organisms in which transposon mutant libraries have been probed in a pooled approach under different stressors include *Shewanella oneidensis* and *Zymomonas mobilis*, assisting the annotation of unknown genes based on their fitness profiles. Therefore, not only is TIS useful for uncovering the effects of chemicals on the organism being studied, but also in probing the system to uncover novel biology.

Observing genome-wide synthetic lethal/sick interactions with antibiotics can also help to generate hypotheses about intrinsic drug resistance mechanisms of bacteria. Knowing that a drug, for instance, has a synthetic lethal or sick interaction with an efflux gene likely points to the antibiotic being pumped from the cell; such as the interaction between *norA* and ciprofloxacin. The interactions with an antibiotic can also give an indication to its mechanism of action if it is not well-characterized or is believed to have multiple targets. It can even be used to probe biological systems in ways that genetic knockouts will not allow.

Collection-based approaches

An extremely high-throughput way to study chemical–genetic interactions is to take an arrayed bacterial collection of single gene deletions or transposon mutants and grow them in the presence of various drugs to see which mutants give a synthetic lethal/sick phenotype with the drug. This concept was pioneered in yeast and works similarly to performing an SGA; however, there is no need to conjugate bacteria to create double mutant strains, as a chemical has replaced one of the genetic perturbations. One study screened the *P. aeruginosa* PA14 transposon mutant library for increased sensitivity to the DNA-gyrase inhibitor ciprofloxacin, a drug commonly used to treat *P. aeruginosa* infections that has decreased in effectiveness due to antibiotic resistance.

The mutants were screened for increased susceptibility to a sub-inhibitory dose (one-half MIC) of ciprofloxacin. Some of the mutants found to be more susceptible to ciprofloxacin were in *ftsK*, as well as DNA replication and repair genes *ruvA*, *recG*, *xerD*, *sss*).

Another study screened the same *P. aeruginosa* mutant library for increased susceptibility to the β–lactam antibiotics imipenem, meropenem and ceftazidime. They discovered 37 mutants with an increased susceptibility to at least one of the β–lactams tested. Interestingly, one of these mutations was in *ftsK*, determined to also increase susceptibility to ciprofloxacin. Other gene disruptions that were synergistic with imipenem included those known to overproduce alginate, an exopolysaccharide present in the mucus of infected cystic fibrosis patients, making this relevant to the treatment of *P. aeruginosa* infections in these patients.

Another organism in which a similar study was performed is *Acinetobacter baylyi*. Gomez and Neyfakh screened the transposon mutant library of *A. baylyi* for hypersusceptibility to one-tenth the MIC of various drugs. Beta-lactam sensitivity was most commonly observed, and some of the strains discovered to be hypersusceptible to β–lactams included those with disruptions in genes encoding efflux components and peptidoglycan biosynthetic genes. Comparably, a characterized transposon mutant library in *M. tuberculosis* was screened for synthetic lethality/sickness with sub-MIC levels of imipenem.

This uncovered 74 gene disruptions that caused sensitivity to the drug, the majority of which were cell wall biosynthetic genes. Discovering these synthetic lethal/sick interactions between a pathogen's genome and clinically relevant antibiotics can prompt further efforts to discover inhibitors of the products of these interacting genes in order to potentiate the chosen antibiotic.

Gene-compound synthetic lethality/sickness has also been widely studied in *E. coli*, as it is a model organism with an extremely well-characterized, widely available, arrayed, single gene deletion Keio collection. Tamae *et al.* studied the synthetic lethality/sickness of the Keio knockouts with sub-inhibitory concentrations of seven different antibiotics and discovered 140 strains sensitive to at least one of the antibiotics.

One study increased the number of antibiotics tested in combination with the Keio deletion collection to 22, discovering 283 strains sensitive to at least one of the drugs. Large-scale phenomic profiling of the Keio collection was also done by Gross and colleagues, who screened the collection in 114 different conditions including many chemical compounds, at varying concentrations.

All of the aforementioned studies made use of colony size on solid media to either qualitatively or quantitatively determine the synthetic lethality/sickness of the gene deletion and drug combination.

Researchers also performed a large-scale chemical genomics study, arraying the Keio collection onto sub-MIC levels of 15 different antibiotics to probe any synthetic lethal/sick interactions. However, in addition to using the standard method whereby endpoint colony size is used to calculate an interaction value using the multiplicative rule, growth rate of the colonies was also tracked and interactions classified based on the maximal growth rate reached.

Growth rate was shown to increase the number of detected synthetic lethal/sick chemical–genetic interactions relative to endpoint biomass measurements alone, from 1052 to 1564 interactions. Notably, deletions in the outer membrane proteins *pal* and *lpoB* were synergistic with antibiotics targeting the cell wall, deletions in LPS biosynthetic genes were synergistic with protein translation inhibitors, and deletions of some exonuclease subunits were synergistic with topoisomerase inhibitors like ciprofloxacin and norfloxacin.

The quantitative nature and high number of chemical–genetic interactions observed when testing dozens of different chemicals across a mutant library enables inferences to be made about drug mechanism of action and gene function at an unprecedented level.

Generating a phenotypic or chemical interaction profile of various deletion strains produces highly useful information. Deciphering which mutants lead to a synthetic lethal/sick phenotype with drugs that either are ineffective or are becoming ineffective due to a rise in resistance allows us to propose potential combination therapies.

Theoretically, if a gene is identified to be synthetic lethal/sick with a drug, it should be possible to either use known inhibitors of that target or screen for inhibitors of that target in order to potentiate the drug (Wright 2012). This would translate from studies of chemical–genetic synthetic interactions to studies of chemical–chemical interactions, aiming to identify drug–drug synergy or drug–adjuvant synergy.

As well, generating a chemical–genomic profile for antibiotics can increase our understanding of their mechanism of action, just as genetic interactions can unveil the function of an unknown gene.

For example, nitrofurantoin is a drug used in the treatment of lower urinary tract infections; however, its exact mechanism of action is not well-understood and is suspected to have a broad range of effects on uropathogenic bacteria. Researchers grew the Keio collection on varying concentrations of nitrofurantoin, and discovered interactions with DNA damage genes, including those involved in the SOS response, as well as nucleotide excision repair.

Clustering genes by their chemical interactions can also reveal information about the functions of uncharacterized genes. For instance, *marR* and *marB* had highly correlated phenotypic profiles in their drug interactions.

Since *marR* is known to repress *marA*, the researchers were led to believe that *marB* performed a similar repressive function, which was supported by follow-up experiments.

Although these arrayed approaches to studying chemical–genetic synthetic lethality/sickness provide a wealth of information, caution must also be exercised when using bacterial collections generated in high throughput, similar to the mentioned caveats of working with large arrayed libraries in the SGA section. Once again, there tends to be a trade-off between the number of conditions that can be explored and whether or not growth kinetics can be considered.

Small molecule screening in deletion backgrounds

Rather than screening the entire genome for gene essentiality under various chemical exposures, it is possible to perform high-throughput small molecule screens against gene deletion strains that have well-characterized synthetic lethal/sick networks. The WTA synthetic lethal genetic network has been described and can be exploited to find molecules that selectively target processes in these networks. Researchers set out to perform a small molecule screen of over 28 000 compounds, looking for those that inhibit growth of the WTA-deficient ”*tarO* strain but not wild-type *S. aureus*.

One of the hits in the screen was amsacrine, an anticancer drug, which had an approximately 16-fold decrease in MIC of the ”*tarO* strain relative to wild type. In order to pinpoint to target of amsacrine within the WTA synthetic lethal network, TIS was performed upon the treatment of wild-type *S. aureus* to the drug.

Through read count depletion and suppressor studies, DltB, involved in the D-alanylation of teichoic acids, was determined to be the target of amsacrine. D-alanylation of LTA is known to be required to maintain viability in the absence of WTA, and thus, knocking out this process results in a synthetic lethal phenotype.

The use of a synthetic lethal/sick approach for drug screening has also been suggested for multidrug resistant *M. tuberculosis* to target parallel α–glucan biosynthetic pathways that have been shown to be synthetic lethal/sick with each other.

Harnessing the information provided by synthetic lethal genetic networks can be used to design smart screens in deletion strains with known inhibitors to target their synthetic lethal/sick partners and achieve a combination therapy for the clinic.

Mechanism of action studies would also be simplified using this approach as the target of the small molecule would be part of the synthetic lethal genetic network, assuming the screened mutant strain is not a hyperpermeable.

Chemical–chemical synthetic lethality

Systematic chemical combinations and potentiation screens

Chemical–chemical combinations are another way to probe for potential synthetic lethal/sick interactions. One of the methodologies used to predict drug interactions is to assume that a combination will be synergistic if the main targets of the two drugs are a synthetic lethal/sick genetic pair.

Antibiotic potentiation screens are one way to broadly assess synergistic or synthetic lethal/sick interactions with chemicals in the bacteria of interest. Generally, the antibiotic of interest is in the screening media at sub-MIC levels, unable cause growth inhibition on its own. Hits may be synergistic or synthetic lethal/sick in combination with the antibiotic of interest.

Typically, hits identified in chemical potentiation screens are verified by exploring the interaction in more detail with a matrix of compound concentrations, often referred to as a 'checkerboard' analysis. Chemical–chemical interactions visualized using a checkerboard can be classified as synergistic, antagonistic or indifferent based on their fractional inhibitory concentration index (FICI).

FICI is calculated as the sum of the MIC of each drug in combination divided by the sum of the MIC of each drug alone. Checkerboard analyses are performed in liquid media in a microtiter plate, exposing the bacterial culture to different combinations of concentrations of two different drugs, which are usually 2-fold diluted between adjacent wells. The cells are grown up to stationary phase under these conditions and growth percentage is reported. Checkerboards showing a 'staircase' phenotype, or an FICI < 0.5 are synergistic.

Several chemical potentiation screens have been performed in the hopes of finding an inhibitor of WTA or other cell envelope components through synthetic lethal/sick interactions with β–lactam antibiotics in methicillin-resistant *S. aureus* (MRSA).

Researchers observed that β–lactam drugs were potentiated in a "*tarO* (WTA-null) MRSA background; therefore, a 2080 compound screen for drugs potentiating the activity of cefuroxime was performed in the hope of finding a WTA inhibitor. Ticlopidine, an antiplatelet drug, was discovered to potentiate cefuroxime and other β–lactams, and the synergy was reversed in the "*tarO* background.

A similar study previously identified tunicamycin as a TarO inhibitor, as it showed immense synergy with β–lactams in MRSA. The synergies between these WTA inhibitors and β–lactams are thought to be due to a synthetic lethal interaction between PBP2 and PBP4. WTA inhibitors indirectly result in improper localization and functioning of PBP4 and when used in combination with β–lactams that target PBP2, this results in a less cross-linked peptidoglycan layer, leading to synthetic lethality.

Another potentiation screen in MRSA identified murgocil, a peptidoglycan inhibitor, to be synergistic in combination with imipenem. The synthetic lethality of this compound combination is likely due to decreased levels of properly localized PBP2 by murgocil, leading to lower levels of β–lactams required to target the PBP.

Chemical potentiation screens have also been performed in Gram-negative bacteria like *E. coli*, not merely Gram positives. Wright and colleagues screened a collection of 30 000 small molecules for potentiation of novobiocin, a Gram-positive drug, in *E. coli*, eliminating those with antibacterial properties alone.

They identified four compounds that were synergistic with novobiocin: echinomycin, A22, pivmecillinam and meropenem. All of these novobiocin adjuvants altered cell shape, likely disrupting influx and efflux mechanisms leading to an accumulation of novobiocin. As well, a minocycline potentiation screen was performed in *E. coli* and *P. aeruginosa* (Ejim *et al.* 2011). One of the hits shown by the screen to be synthetic lethal/sick with minocycline was loperamide, an anti-diarrheal drug.

Downstream checkerboard experiments confirmed the synergy of loperamide with minocycline and other tetracycline antibiotics. The synthetic lethality/sickness was found to be due to their combined interactions with the proton motive force (PMF). Loperamide was found to dissipate the electrical component of PMF, resulting in an increase in the chemical component of PMF, driving higher levels of tetracyclines into the cell.

Chemical–chemical synthetic lethality screens can also provide information about mechanism of action. For example, a chemical screen for growth inhibition of *E. coli* MC1061 identified 186 bioactive compounds with unknown mechanisms of action.

Bioactives were combined with 14 antibiotics with known mechanisms of action at one-quarter and one-eighth MIC to assay for synergy; 45 of the compounds were found to be synergistic with the known antibiotics. Synergies can lend information about the mechanism of action of unknown compounds—for example, MAC-0003199 showed synergy with norfloxacin, a DNA gyrase inhibitor, and was confirmed to be a DNA gyrase inhibitor itself. As well, molecules with similar structures were shown to cluster by synergy patterns.

Using these methods to look for chemical–chemical synergies or synthetic lethality/sickness provides the ability to intelligently identify combination drug treatments. Identifying adjuvants or synergizing antibiotics for a drug can increase its efficacy in strains that are normally unaffected by treatment. Dose sparing is another reason to screen for efficacious drug combinations; if a drug can be potentiated by another, lower doses can be used of each, thus increasing the therapeutic window for compounds that may exhibit host toxicity at higher concentrations.

A drug's chemical–chemical interaction profile can also be used as a tool for determining the mechanism of action. By comparing it to profiles of drugs with known mechanisms, we can find those that cluster together and infer that the drugs act in a similar manner in the cell. Drugs of the same class tend to produce similar chemical–chemical interaction profiles. As well, the drugs observed to have a synergistic interaction when used in combination might target genetic synthetic lethal/sick pathways. This provides a 'one-two punch' to sufficiently disrupt or inactivate an essential pathway for cell viability, assuming that the drugs synergize due to their primary targets, and has been used to predict synergies.

Understanding synthetic lethal chemical combinations can also assist in adjusting treatment courses for patients whose infections evolve to become resistant to one of the drugs in an administered combination. Another similar synergistic drug can be substituted in the combination for the drug that no longer works, assuming no cross-resistance is exhibited.

12

QUANTITATIVE INHERITANCE

An inheritance of a character that depends upon the cumulative action of many genes, each of which produces only a small effect. Examples of such quantitative characters include spore production in ferns, height of trees, and nectar production in buttercups. Usually the character shows continuous variation, i.e. a gradation from one extreme to the other.

QUANTITATIVE GENETICS

The analysis of traits whose variation is determined by both a number of genes and environmental factors.

Polygenic inheritance is the additive effects of two or more genes on a single phenotypic characteristic.

- It is important not to confuse multiple alleles with polygenic inheritance.
- Blood type is an example of multiple alleles, whereby blood type is controlled by three possible alleles) all occurring at one locus. Polygenic inheritance involves many genes at several different loci.

Continuous vs Non-continuous Variation: Variation is expressed by the different phenotypes for a characteristic. Variation can be:

Continuous. Exhibits a wide range of phenotypes ranging from one extreme to the other.

Discontinuous. Usually only has 2 forms.

Polygenic Inheritance can contribute to continuous variation.

Good examples of quantitative inheritance are afforded by:

Kernel Color in Wheat

The first experiment on polygenic inheritance was demonstrated by Swedish Geneticist H.Nilsson - Ehle (1909) in wheat kernels. Kernel colour is controlled by two genes each withtwo alleles, one with red kernel colour was dominant to white. He crossed the two purebreeding wheat varieties dark red and a white. Dark red genotypes R1R1R2R2 and whitegenotypes are r1r1r2r2. In the F1 generation medium red were obtained with the genotypeR1r1R2r2. F1 wheat plant produces four types of gametes R1R2, R1r2, r1R2, r1r2. Theintensity of the red colour is determined by the number of R genes in the F2 generation.

Skin Color in Man

The characters that Mendel studied are sometimes referred to as discrete characters becausethey can only be classified on an "either-or" basis (e.g., purple or white flowers, green or yellowseeds). Many characters cannot be classified in this manner because they vary in a populationacross a continuum (gradient). For example, the figure blow illustrates that skin color inhumans is a quantitative character. Quantitative characters usually indicate that the characteris controlled by more than one gene (polygenic inheritance). A simplification of the geneticsof skin color in humans shows that three genes interact to determine the level of pigment in anindividual's skin.

The dominant alleles (A, B, and C) each contribute one "unit" of pigment tothe individual, and their effects are cumulative such that individuals with more of these alleleswill be darker than those with fewer alleles. The recessive alleles (a, b, and c) do not contributeany units of pigment. Therefore, skin color is related to the number of dominant alleles presentin each individual's genotype. A cross of two completely heterozygous parents produces sevenphenotypes in their offspring, ranging from very light to very dark skin. The distribution of skin color in the offspring would resemble a bell-shaped curve because there would be moreindividuals with intermediate skin colors than either extreme. As the number of genes involved increases, the differences between the various genotypes become more subtle and the distribution fits the curve more closely.

Some of the major differences between monogenes and polygenes in inheritance are as follows:

Monogenes/Monogenic Inheritance

- They produce discontinuous variations in the expression of traits.
- A single dominant allele expresses the complete trait.
- Monogenic inheritance controls qualitative traits.
- A character is represented in an individual by a pair of alleles
- F1 individuals are similar to dominant parent.
- F2 individuals resemble both the parents in the ratio of 3: 1.
- No intermediates are produced in monogenic or qualitative inheritance.
- There is no cumulative action in the presence of two dominant genes

- Individuals with dominant phenotype are more numerous than with recessive phenotype.

Polygenes/Polygenic Inheritance

- Polygenes produce continuous variations in the expression of traits.
- A single dominant allele expresses only a unit of the trait.
- Polygenic inheritance controls quantitative or metric trait
- A character is represented by one to several pairs of alleles.
- F1, individuals are intermediate between the two parents.
- Depending upon the number of polygenes, 2/4 (one pair), 2/16 (two pairs) or 2/64 (three pairs) F: individuals resemble the parental types.
- Intermediates are quite common in polygenic or quantitative inheritance.
- The dominant genes have cumulative effect on the expression of the trait.
- Individuals with dominant trait are usually as few as with recessive trait. Intermediate forms are more numerous

Typical monohybrid cross of Mendel involving a monogenic trait shows the following characteristics

1. The parents belong to two distinct phenotypic classes: homozygous dominant and homozygous recessive
2. All F_1 offspring resemble one of the parents because one gene is dominant over the other.
3. 3. F_2 offspring show segregation into dominant and recessive phenotypes in the ratio of 3:1

Monohybrid cross involving a polygenic trait shows the following characteristics:

1. The parents here too belong to two distinct phenotypic classes: homozygous dominant and homozygous recessive.
2. All F1 offspring show a trait intermediate between those between those of parents because the dominant genes fail to fully suppress their recessive alleles.
3. F2 offspring show a wide range of intermediate condition and very limited parental traits.

QUANTITATIVE TRAITS

- The quantitative traits are governed by many genes at different loci and each contributes such a small amount to the phenotype that their individual effects can not be detected by Mendelian methods.
  - Each gene exerts a small additive effect.
  - Due to incorporation of more than one gene in expression of a given trait, quantitative traits are also referred as polygenic traits.

- There is no dominance .
- There is no epistasis or interaction among the different loci contributing to the value of the trait.
- The continuous variation in quantitative traits results from the combined effects of genetic and environmental variations. Neither factor governs the variation exclusively. Due to this, It is also called as multifactorial or complex trait.

Not All Polygenic Traits Show Continuous variation

- In addition to quantitative traits, there are two classes of polygenic traits which do not show continuous variation.
  1. Meristic TraitsIn this phenotype is described by whole numbers, integral categories, discontinuous.

     For Ex- No. of seeds in a pod or the number of eggs laid by a chicken in a year,

Litter size in cats, tongue rolling.

- These are quantitative traits, but they do not have an infinite range of phenotypes: for example, a pod may contain 2, 4, or 6 seeds, but not 5.75

Threshold Traits

- These are polygenic (and frequently multifactorial), but they are distinguished from continuous and meristic traits by having a small number of discrete phenotypic classes.
- It has an underlying quantitative distribution, but the trait appears only if a threshold is crossed.
- Only those individuals exceeding thethreshold on the liability scale will expressthe trait
- Example is Type II diabetes, A population can be divided into just two phenotypic classes for this trait— individuals who have Type II diabetes and those who do not— so at first glance, it may appear to more closely resemble a simple monogenic trait. However, no single adult-onset diabetes gene has been identified. Instead, the combination of alleles present at multiple contributing loci gives an individual a greater or lesser likelihood of developing the disease.

The Multiple-Gene Hypothesis for Quantitative Inheritance

- By experimental output from Nilsson-Ehle (1999)
- Parental type with Red and White grain colorhas given intermediate color in F1.
- F2 has given 5 color classes, from red to white were distinguishable in a ratio of 1:4:6:4:1.

- This result is explained by assuming that two loci are assorting independently each with two alleles, such that one potential additive allele produces a unit of red color whereas the one potential non-additive allele does not. SO we can envision the generation of such variations.
- Each of the F2 plants has 4, 3, 2, 1, or 0 additive alleles. F2 plants with no additive alleles are white (aabb) while with 4 additive alleles are red (AABB). Plants with 3, 2, or 1 additive alleles constitute the other three categories of red color observed in the F2 generation.
- The greater the number of additive alleles in the genotype, the more intense the red colorexpressed in the phenotype.

The **Multiple-gene hypothesis** consists of the following major points:

1. Phenotypic traits showing continuous variation can be quantified by measuring, weighing, counting, and so on.
2. Two or more gene loci, often scattered throughout the genome, account for the hereditary influence on the phenotype in an additive way. Because many genes may be involved, inheritance of this type is called polygenic.
3. Each gene locus may be occupied by either an additive allele, which contributes a constant amount to the phenotype, or a nonadditive allele, which does not contribute quantitatively to the phenotype.
4. The contribution to the phenotype of each additive allele, though often small, is approximately equal. While we now know this is not always true, we have made this assumption in the above discussion.
5. Together, the additive alleles contributing to a single quantitative character produce substantial phenotypic variation

QUANTITATIVE GENETICS IN PLANT BREEDING

Introduction

For almost one century scientists have dissected the genetic architecture of quantitative traits in plants using Quantitative trait loci (QTL) analysis. These analyses establish associations between genetic markers and the phenotypic variation of a quantitative trait in a segregating population. The techniques used to obtain markers and physiological phenotypes have been constantly improved through history. Recently, the price drop of high-throughput technologies have allowed plant researchers to quantify the general abundance of transcripts, proteins, or metabolites in segregating populations.

These studies show that there are multiple benefits in using "omic" technologies for QTL analyses, even when the goal is to characterize physiological phenotypic diversity. First, molecular phenotypes are the initial step toward the production of physiological phenotypes and its regulation underlies much of phenotypic diversity. Second, the availability of genome-wide information significantly increases the ability to identify candidate genes for QTLs. Third, molecular traits measured at system scale allow estimation of the effect of QTLs in the genetic pathways of interest, or identification of additional gene networks altered by the loci responsible for the variation.

Finally, molecular traits offer researchers a better understanding of how mutation drives physiological variation and what are the evolutionary forces acting at primary levels.

High-throughput sequencing, or HTS, allows the rapid and cost–effective generation of massive amounts of short sequences or reads (Metzker, 2010). The potential of this technology for mapping loci responsible for phenotypic differences in plants has already been demonstrated by identifying genes containing EMS-induced mutations in samples of pooled F2 individuals. HTS technologies have been in the market for a few years, and new methods are being developed that will be cheaper, require less sample processing, and will produce more and longer reads.

It is therefore clear that very soon HTS will be the tool of choice for QTL analyses. One important limiting factor remains to be eliminated: Data analysis. It requires long and computationally intensive pipelines that need to be customized for each particular experimental set up. An increasing number of new algorithms are constantly released to the community, and the debate on which pipelines return the most accurate results is still ongoing. Comparing, combining, and customizing these pipelines requires simple Unix or Linux commands and greatly benefits from knowledge in powerful statistical software such as R, and in scripting languages, such as Perl or Python.

For non-bioinformaticians, integrated solutions with convenient interfaces are becoming popular both from collaborative open projects and companies. A popular website that keeps an actualized list of the available software tools is www.seqanswers.com, where users and developers also discuss new technological advances and pipelines. In terms of the computational equipment required for HTS data analysis, the majority of tools are developed for Linux or Unix based systems.

Although parts of the analysis can be performed in any modern computer, machines with dozens of gigabytes of RAM are recommended in cases where reference sequences form the species considered are available, or with hundreds if no reference exists. An alternative option that is likely to become popular is to rent storage and computing power in specialized centers, or "the cloud".

Due to the fast improvement of HTS, this review intends only to capture a snapshot in time of the possibilities that it offers for molecular marker discovery, genotyping, and molecular phenotyping in segregating populations of plants. This review has the purpose of helping researchers who have not incorporated this technology to their work to think about the requirements and possibilities of HTS.

By no means this review refers to all available experimental designs or analysis tools, and the solutions proposed here are mere suggestions that will certainly soon be substituted by new and better ones. A guide map of the methods proposed in this review is depicted in Fig. 1.

Library Preparation

Sample preparation protocols are continuously improved to use fewer amounts of biological material, be completed faster, and reduce the bias in their output. As an example, most current protocols allow multiplexing samples by adding a short sequence tag to all reads in a library, a convenient feature given the increasing numbers of reads produced per HTS run.

The same companies that developed the HTS sequencers commercialize library preparation protocols optimized for the most common experimental designs. There are also kits from other companies that give comparable results and may be more cost efficient.

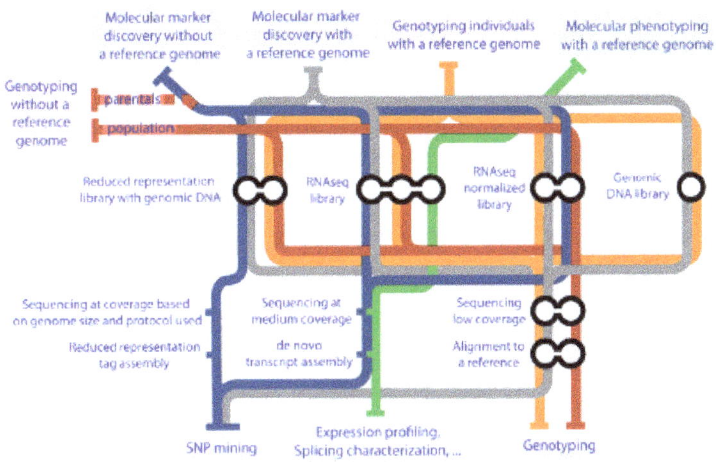

Fig. 1. Guide map to the proposed pipelines for SNP identification, genotyping, and molecular phenotyping for QTL analysis in plants.

Finally, many researchers are developing custom protocols to obtain specific information such as the transcribed strand in RNA-seq experiments, the rate of RNA degradation, or the positions occupied by RNA polymerases, just to name a few.

Quality Control and Pre-Processing

Assessing the quality of HTS reads includes detection of biases on base composition, base quality, and sample complexity. The quality of the sequences has an impact on the reliability of the biological interpretations resulting from the analysis. Part of these biases are introduced by the sample preparation protocols, particularly during cDNA synthesis in RNA-seq experiments and PCR amplification. Additional biases are particular to each HTS technology or specific to each run of the sequencers.

After quality control it is usually necessary to pre-process the reads by trimming low quality nucleotides and adapter sequences. At this stage, foreign sequences such as vectors or DNA from organisms contaminating the samples can also be removed. Depending of the type of libraries sequenced further pre-processing may be needed, such as trimming poly A or poly T tails and terminal transferase tails in RNA-seq libraries. In cases where several libraries have been multiplexed, reads should be separated by their barcode.

Both quality control and pre-processing can be easily performed with basic scripts written in Perl (Bioperl), R (Bioconductor), or Python. For non-programmers, there are some convenient tools that can carry out all or some of these tasks.

Molecular Marker Discovery

Depending on the availability of a reference sequence short reads will be aligned or *de novo* assembled using one of the multiple tools available. There are a number of recent articles that compare the most popular algorithms and software available for these purposes. Please note that the methods proposed below are directed to developing molecular markers for QTL analysis and not to identify the mutation underlying the QTL, which requires much deeper sequencing.

With a Reference Sequence

A cost efficient solution to obtain molecular markers is to sequence DNA or RNA from the parental genotypes and mine polymorphisms from the resulting reads. These polymorphisms can be used later to design PCR markers or a high-throughput genotyping assay for the full population.

This approach works remarkably well in diploid and polyploidy species using as low an amount of sequence as 5× coverage, meaning five times the size of the genome under study.

A recent article reviews the methods and tools available for single nucleotide polymorphism (SNP) identification and genotyping. To align the reads to the reference, mapping softwares based in "seed methods" are preferred despite their slower nature because their robustness to polymorphisms. Before SNP calling users may consider removal of the reads that map to multiple locations in the reference, and of duplicated reads that may have been generated from PCR artifacts. A recent pipeline also recalibrates the quality of the nucleotides in the reads to correct for the high error rates in HTS, and realigns reads in complex genomic positions where the fast processing alignment algorithms may have failed.

Commonly used indicators of the veracity of polymorphisms are based in the amount and quality of reads showing the polymorphism, frequency of the observed alleles, quality of the alignment, and/or proximity to other polymorphisms. There are some basic and popular options for calling polymorphisms from aligned reads, tools specialized in the analysis of reads from particular sequencing platforms, that have the ability to detect structural variation, or that have into account the quality of the reference in addition to the quality of the reads. An essential method to control for the quality of the data analysis process is visual inspection through genome viewers specialized in HTS datasets.

Without a Reference Sequence

High-throughput sequencing sequences can serve to construct the necessary reference to identify molecular markers if it is not already available. Although assembling *de novo* a complete genome sequence is possible with HTS, it requires very deep sequencing and extensive bioinformatic analysis, even more given the relatively large size of most plant genomes. A more efficient option is sequencing mRNA, which greatly reduces sample complexity in comparison with genome sequencing and has the advantage of offering functional information such as coding polymorphisms or expression levels.

A comprehensive compilation of the methods and tools available for transcriptome assembly has been recently published. *De novo* assembly algorithms greatly benefit from long and paired-end reads, but are extremely sensitive to errors and polymorphisms and will not perform well during assembly of datasets from mixed genotypes or highly heterozygous individuals. The amount of new genomic positions detected in RNA-seq experiments decrease exponentially as the number of reads increases. The majority of medium and highly expressed transcripts in a sample are detected at low coverage, and increasing coverage will mainly add non-coding RNAs and low expressed transcripts at a very high cost.

If the objective is to assemble complete transcriptomes, obtaining samples from diverse tissues, time points, and conditions is preferred to depth of sequencing. Even in the best possible conditions assemblies from RNA-seq reads will return only a subset of the existing transcripts, many of which will be fragmented.

This is expected due to low expression of particular transcripts, the non-uniform read coverage, and the presence of different isoforms per gene. To help assembly of low expressed transcripts researchers can use normalization protocols that deplete the most abundant transcripts from the samples (Christodoulou et al., 2011). In any case, contigs resulting from *de novo* assembly can be effectively used as a reference for molecular marker detection and characterization of transcripts in un-sequenced genomes.

When highly similar genotypes are compared, RNA-seq may not be the best option since it mostly targets coding regions, which are less diverse than non-coding regions. In these cases researchers can construct reduced representation libraries by shearing DNA using restriction endonucleases and size-selecting the fragments that will be sequenced. Reads from these libraries can be clustered by similarity and mined for polymorphisms close to the restriction sites; or used to detect the presence–absence of particular tags, indicating a polymorphism in the restriction site itself.

Obtaining polymorphisms from reduced representation libraries is more efficient when a reference sequence is available. However, researchers have already developed tools to genotype samples from these tags using a low number of reads from organisms without a reference, or to reconstruct part of the targeted genome using paired-end sequencing. Additional protocols to obtain markers from reduced representation libraries exist in which different combination of restriction enzymes are used for each of the genotypes involved, or that do not shear the DNA but filter the reads for single copy sequences. The amount of reads necessary to perform this type of analysis depends on the size of the genome, the restriction enzymes used, and the availability of a reference.

Genotyping Populations

With the price drop of the HTS technologies and the possibility of multiplexing samples, genotyping an entire population has become realistic. In the case of a sequenced system such as rice, generating reads from the individuals of a population at 0.02–0.055× coverage allowed high-density genotyping by comparisons with the parental genotypes, or by inferring the parental genotypes from the polymorphisms found in the population.

Since erroneous polymorphism calls are expected at low coverage, more or less complex algorithms need to be defined to correctly genotype each polymorphism in each individual. In addition, a reference sequence can serve researchers to design enrichment essays that will target their preferred genomic locations, although at high cost.

Molecular Phenotyping

The list of molecular phenotypes that can be quantified with HTS is extensive and is rapidly increasing. Examples of these phenotypes are protein–RNA interactions, translation rates, transcription rates, protein–DNA interactions, RNA degradation rates, RNA secondary structure, transcription start positions, chromatin accessibility, methylation states, natural antisense transcription or small RNA profiles. QTL analysis using these phenotypes as traits is an exciting field that remains un-explored. Therefore, the computational frameworks to quantitatively compare these phenotypes between individuals will need to be established.

EXPRESSION PROFILING WITH HTS

Although many cases of phenotypic variation caused by coding polymorphisms have been documented, variation in gene expression has been shown to underlie much of phenotypic diversity. One method to detect differences in expression between individuals using HTS is to sequence 26–27 nucleotide-long tags from expressed transcripts. A recent study shows that this method reaches saturation in mice with 6–8 million reads per sample.

Its advantages over sequencing full transcripts are the lower cost, higher sensitivity, reduced bias during amplification due to the fixed fragment lengths, and use of simplified statistical models to calculate differential expression. On the other hand, methods based in tags will not detect the majority of coding polymorphisms and isoforms, and require a close enough reference sequence to extract biologically meaningful results.

RNA-seq is rapidly becoming a standard in expression profiling because of its simple protocol of preparation, digital nature, large dynamic range, and high sensitivity in comparison with previous technologies. In addition, it can serve to genotype individuals, identify novel transcripts, characterize alternative splicing, and quantify allele specific expression.

Due to the novelty of the technique there is no consensus on which sample preparation protocols present fewer biases. However, strand-specific methods could become a standard because of their increased precision due to their ability to distinguish between sense and antisense transcripts. In terms of experimental designs, it is necessary to randomize and replicate biological samples, as with any other type of genome-wide analysis. There is little consensus about the depth of sequence needed for expression profiling with RNA-seq.

Recent estimates range between 30 million reads to compare the expression profiles of two samples, to 100 million reads to detect most transcribed genes and quantify isoforms, to 500 million to obtain accurate profiles, including low expressed transcripts. In any case, it is advisable to balance the number of reads between samples in the same experiment in order to perform accurate expression comparisons.

Expression profiling from HTS datasets is necessarily based on counting the reads mapped to each transcript in a reference sequence. When a reference genome or transcriptome is not available, it can be reconstructed using *de novo* assembly of the reads for at least one of the genotypes as described above. The simpler and less computational intensive protocol for expression profiling is to map the RNA-seq reads to known (or *de novo* assembled) transcripts and a set of possible exon–exon junctions (when available) to detect alternative splicing.

However, in organisms with sequenced genomes this protocol will not allow detection of novel exons, transcripts, and isoforms. The preferred pipeline involves aligning the reads to the genomic reference using an alignment tool that splices the reads to detect intron–exon junctions.

A challenge for expression analyses in samples from two unrelated individuals is the need to perform robust quantification of reads generated from two or more alleles. This implies that reads with the closer genotype to the reference will align better than reads from a more distant genotype, in which more polymorphisms may interfere with their ability to map. In these cases, aligners based in seed methods will perform better than those based in the Burrows–Wheeler Transform algorithm.

Although most studies ignore this problem, there are solutions that go from identifying and removing the polymorphisms that cause these biases, aligning the reads to all references from the genotypes involved or including the polymorphisms found in the references. When two references are used, a potential problem may arise from motifs that are more abundant in one reference with respect to the other if only uniquely mapped reads are counted.

The use of longer reads and/or paired-end reads greatly decreases the number of ambiguously mapped reads. In addition, there are robust methods to assign these multi-mapped reads to a single location. There are a number of tools to count the number of reads aligned to each transcriptional unit to calculate expression, most of which require knowledge of Perl, Phyton, Linux/Unix, or R.

Some alignment tools can directly calculate the number of reads per transcript and/or a measure of expression based in the reads (or fragments) per gene size in kilobases per million reads mapped, called RPKM (or FPKM). However, these expression units show biases depending on the length, number, abundance of the transcripts present in the samples, or because of technical replication. For this reason researchers have developed dedicated R/Bioconductor packages to calculate differential expression between samples based on raw read counts per transcript. In addition, there are software packages that take into consideration the biases inherent to RNA-seq when calculating expression or performing downstream analyses such as gene ontology over-representation studies.

High-throughput sequencing datasets allow quantification of expression for each isoform separately, resulting in significantly more accurate estimates than calculating expression at the gene level. For this, users must first identify splicing events from the reads that align to exon–exon junctions. Quantifying isoform expression is complicated since most reads in an alternatively spliced transcript cannot be assigned to a single isoform. The most promising methods to address this complex problem take advantage from the information offered by paired-end and/or unambiguously mapped reads. One advantage of going through the intricate process of identification of alternative splicing is that it can also be used as a trait for QTL analysis.

Allele Specific Expression in Hybrids

An alternative to sequencing a full segregating population to perform eQTL analyses is to sequence F_1 hybrid individuals, where allele specific expression can be calculated for loci with coding polymorphisms. For any gene, both alleles in the hybrid share the same cellular environment and, as a result, changes in expression between alleles must necessarily be due to *cis*-acting regulators (Cowles et al., 2002). Trans-acting eQTLs can be inferred by performing RNA-seq in the parentals and comparing the differences in expression levels between alleles in the hybrid with the differences between the parentals (Wittkopp et al., 2004).

Despite the considerable reduction in price and simplicity of experimental design, this method has several drawbacks. Allele specific expression can only be calculated in transcripts with coding polymorphisms that are highly covered, and it is very dependent on read and transcript length.

New statistical approaches are being developed that will overcome these limitations, starting by being able to estimate false discovery rates and allele specific alternative splicing.

In summary, HTS is changing the way we perform QTL analysis by allowing high-throughput genotyping of populations and phenotyping of traits with a precision not achievable before. It is clear that HTS has not reached its peak of development, and that tools and algorithms will have to be modified according to the new technological improvements. Nevertheless, the first experiments using this technology have already identified exciting possibilities for the characterization of natural variation in plants.

13

MULTIPLE ALLELES (BASED ON CLASSICAL CONCEPT OF ALLELOMORPHISM)

Allele, also called allelomorph, any one of two or more genes that may occur alternatively at a given site (locus) on a chromosome. Alleles may occur in pairs, or there may be multiple alleles affecting the expression (phenotype) of a particular trait. The combination of alleles that an organism carries constitutes its genotype. If the paired alleles are the same, the organism's genotype is said to be homozygous for that trait; if they are different, the organism's genotype is heterozygous. A dominant allele will override the traits of a recessive allele in a heterozygous pairing. In some traits, however, alleles may be codominant—i.e., neither acts as dominant or recessive. An example is the human ABO blood group system; persons with type AB blood have one allele for A and one for B. (Persons with neither are type O.)

Most traits are determined by more than two alleles. Multiple forms of the allele may exist, though only two will attach to the designated gene site during meiosis. Also, some traits are controlled by two or more gene sites. Both possibilities multiply the number of alleles involved. All genetic traits are the result of the interactions of alleles. Mutation, crossing over, and environmental conditions selectively change the frequency of phenotypes (and thus their alleles) within a population. For example, alleles that are carried by individuals with high fitness (meaning they successfully reproduce and pass their genes to their offspring) have a higher likelihood of persisting in a population than alleles carried by less-fit individuals, which are gradually lost from the population over time.

FEATURES OF ALLELES

Features of multiple alleles

- The same genes have more than two alleles.
- All multiple alleles in homologous chromosomes occupy the respective loci.

- A chromosome or gamete only has one group allele.
- Each human contains only two separate gene alleles, one for each homologous pair of chromosomes carrying the gene.
- Multiple alleles express various alternatives of one trait.
- Different alleles can exhibit codominance, dominance-recessive behaviour or incomplete dominance.
- Multiple alleles validate the Mendelian sequence of inheritance.

Examples of multiple alleles

The most common examples of multiple alleles are

- The coat colour of rabbits,
- A, B, AB and O blood groups in humans
- The eye colour in Drosophila

Mendel implied that only two alleles, one dominant and one recessive, could exist for a given gene. We now know that this is an oversimplification. Although individual humans (and all diploid organisms) can only have two alleles for a given gene, multiple alleles may exist at the population level such that many combinations of two alleles are observed. Note that when many alleles exist for the same gene, the convention is to denote the most common phenotype or genotype among wild animals as the wild type (often abbreviated "+"); this is considered the standard or norm. All other phenotypes or genotypes are considered variants of this standard, meaning that they deviate from the wild type.

The variant may be recessive or dominant to the wild-type allele. An example of multiple alleles is coat color in rabbits (Fig. 1). Here, four alleles exist for the *c* gene. The wild-type version, C^+C^+, is expressed as brown fur. The chinchilla phenotype, $c^{ch}c^{ch}$, is expressed as black-tipped white fur. The Himalayan phenotype, c^hc^h, has black fur on the extremities and white fur elsewhere.

Finally, the albino, or "colorless" phenotype, *cc*, is expressed as white fur. In cases of multiple alleles, dominance hierarchies can exist. In this case, the wild-type allele is dominant over all the others, chinchilla is incompletely dominant over Himalayan and albino, and Himalayan is dominant over albino. This hierarchy, or allelic series, was revealed by observing the phenotypes of each possible heterozygote offspring.

The complete dominance of a wild-type phenotype over all other mutants often occurs as an effect of "dosage" of a specific gene product, such that the wild-type allele supplies the correct amount of gene product whereas the mutant alleles cannot. For the allelic series in rabbits, the wild-type allele may supply a given dosage of fur pigment, whereas the mutants supply a lesser dosage or none at all. Interestingly, the Himalayan phenotype is the result of an allele that produces a temperature-sensitive gene product that only produces pigment in the cooler extremities of the rabbit's body.

Fig. 1. Four different alleles exist for the rabbit coat color (C) gene.

Alternatively, one mutant allele can be dominant over all other phenotypes, including the wild type. This may occur when the mutant allele somehow interferes with the genetic message so that even a heterozygote with one wild-type allele copy expresses the mutant phenotype. One way in which the mutant allele can interfere is by enhancing the function of the wild-type gene product or changing its distribution in the body.

One example of this is the *Antennapedia* mutation in *Drosophila*. In this case, the mutant allele expands the distribution of the gene product, and as a result, the *Antennapedia* heterozygote develops legs on its head where its antennae should be.

MULTIPLE ALLELES AND ISOALLELES

Mendel's concept of inheritance envisaged alternative forms of same gene or factor. This would mean that each gene had two alternative forms or allelomorphs, one being dominant or partially dominant and the other recessive; one being wild form and the other mutant. If the mutant form developed from wild form due to mutation, there is no reason, why we should not expect that the wild form can mutate in more than one ways. The mutant form can also mutate once again to give rise to another mutant form. Therefore, a gene can have more than two allelomorphs. These allelomorphs make a series ef multiple alleles. Relationship between different alleles may or may not be a simple dominant-recessive relationship, but they are usually characterized by their location on the same locus, and therefore, by lack of crossing over between them. However, discovery of pseudoalleles has rendered this definition out of date, that multiple alleles, can occupy different sites within a gene and that crossing over between them can take place.

It is also possible that several alleles may exhibit themselves within the same phenotypic range, whether wild or mutant. Such allelesexpressing themselves within the same phenotypic range are called isoalleles. If the phenotype is wild, these are normal or wild isoalleles and if the phenotype is mutant, they will be called mutant isoalleles. Such examples are many and one of them is noticed in eye colour in *Drosophila,* where several alleles exhibit red eye colour (W^{+s}, W^{+c}, W^{+g}). The phenomenon of multiple allelism will be illustrated in this section with the help of a number of examples, both from animals and plants.

MULTIPLE ALLELES IN FRUIT FLIES

In 2000, scientists completed the complex genome mapping of the common fruit fly, *Drosophila melanogaster*. The fruit fly has been and continues to be a valuable laboratory animal due to its high reproduction rate and the ease with which large numbers of flies can be kept and analyzed. At approximately 165 million base pairs, a fruit fly's DNA is significantly smaller than that of a human. While humans have 23 chromosomes, fruit flies have only four. Nonetheless, there are approximately 17,000 genes on only four chromosomes. Each gene is responsible for a distinct aspect of the fly and is subject to mutation and the emergence of new alleles.

Normally, the wings of *Drosophila melanogaster* are quite long. Two mutations occurred at the same locus in different flies' one of which resulted in vestigial (reduced) wings and the other in antlered (less developed) wings.

When a fly with vestigial wings is crossed with another with antlered wings, the resulting F1 hybrids have intermediate wing lengths, indicating that none of the mutated genes is dominant. Occasionally referred to as the vestigial antlered compound, this hybrid contains two mutated genes at the same locus. Mendelian segregation and recombination are evident.

Apart from the vestigial and antlered wings, there are other phenotypes: *nicked wings, strap wings,* or *no wings* at all. The gene variants responsible for these phenotypes are the multiple alleles in this particular fruit fly population.

MULTIPLE ALLELES AMONG HUMANS

There are traits in humans and other organisms that have three or more different types of alleles (genes). When a trait has three or more distinct alleles, we refer to it as having multiple alleles inheritance. The human ABO blood type alleles/trait is an example of a trait with multiple alleles. Three distinct alleles exist: allele A (I^A), allele B (I^B), and allele i (I^O or i).

If the allele A is present on the chromosome, protein A is produced, and the red blood cells of that individual contain protein A on their membrane. If the chromosome contains the allele B, protein B will be produced, and the red blood cells' membranes will contain protein B. Finally, if allele i is present on the chromosome, neither protein A nor protein B will be synthesized. These three alleles make up the ABO blood group trait.

The inheritance pattern of allele A and allele B shows codominance (co-dominance). Co-dominance occurs when neither allele is dominant over the other, and a heterozygous individual expresses both phenotypes. That is, if an individual possesses allele A on one chromosome and allele B on the second homologous chromosome, both proteins are expressed and red blood cells contain both protein A and protein B on their cell membranes.

The ABO blood type hereditary scheme in human beings is an illustration of multiple alleles blood types. Phenotypes are classified into four categories: blood group A, blood group B, blood group AB, and blood group O. In this case, the population has three alleles. The I^A allele when expressed results in the presence of A molecules on red blood cells, the I^B allele expression leads to the presence of B molecules on red blood cells, and the I^O allele expression results in the absence of such antigens on the red blood cells. The I^A and I^B alleles are not only codominant but both of them are dominant over the I^O allele. I^O allele is recessive and thus will be expressed if I^A or I^B is absent.

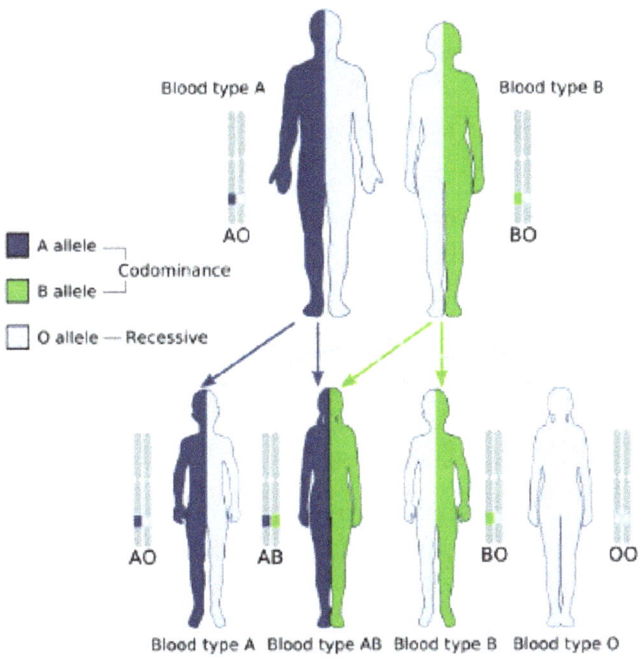

Fig. 2: Possible blood types of offspring from a cross between male and female parent with blood type A and Blood type B, respectively.

Although a population contains three alleles, each individual inherits only two of them from their parents. This results in the genotypes and phenotypes depicted below. Consider that when three alleles are present, there are six distinct genotypes. The number of phenotypes possible is determined by the dominance relationships between the three alleles. Watch the video below to understand multiple alleles in the form of ABO Blood types.

MULTIPLE ALLELES IN PLANTS

While it is widely believed that potato tuber shape is continuous, visual phenotypes such as round or long tubers can be discerned at the diploid level. Although experimental evidence for the presence of multiple allele systems for a potato tuber is being reported for the first time, this study in potato can be compared to one in maize. The recessive allele for tuber shape can be regarded as the qualitatively recognized null or near-null allele. Variation between dominant alleles is quantitative. The view that the (most) recessive allele can be considered a null or near-null allele is consistent with how quantitative effects at a multiallelic locus are described. When additional metric traits are resolved into Mendelian factors using heterozygous parents in experimental designs, conclusions can be drawn about the relative importance of multiple alleles traits to multiple loci in explaining quantitative genetic alleles variation.

MULTIPLE ALLELES IN BACTERIA

Bacteria possess a large number of genes, several of which have multiple alleles. These various wild-type alleles are frequently associated with distinct types of virulence and can be used to classify subspecies (e.g., housekeeping genes for Multi Locus Sequence Typing, MLST). As a result, it is critical to rapidly identify not only the desired gene but also the relevant allele.

Currently available sequencing-based methods are limited to mapping reads to each known allele reference, which is a lengthy process. Understanding and predicting the pathogenic impact and outbreak potential of a bacterial infection requires more than knowledge of the species responsible for the infection.

Bacterial virulence is frequently regulated on a sub-species level by a collection of specific genes or even alleles, necessitating the use of distinct treatment strategies for infections caused by the same bacterial species.

Antibiotic resistance, for example, is a well-known example of how minor variations in a gene can result in a diverse collection of antibiotic resistance profiles within a single taxonomic group.

Additionally, different alleles of the same gene may be responsible for distinct adhesion and invasion strategies, responses to the infected organism's immune response, and toxin production.

Apart from its importance for understanding virulence, identifying alleles of specific genes contributes to a more precise classification of bacteria.

MULTIPLE ALLELISM, PLEIOTROPY AND EPISTASIS

Multiple allelism may be misconstrued with other common terms in Genetics, such as pleiotropy and epistasis. Let us understand the difference between multiple allelism (defined as the condition of having multiple alleles) and other genetic conditions.

- Pleiotropy refers to the condition wherein more than one gene can have multiple effects on the phenotype. While multiple allelism involves only a single gene with many variants (referred to as multiple alleles), pleiotropy entails more than one gene that determines phenotype. For example, a higher frequency of individuals with albino would have cross-eyes than the pigmented individuals. Thus, apart from having no sufficient pigment production in their skin and hair, another possible feature of albinoes is a cross-eyed trait. However, not all albinoes show this trait indicating that in such cases, the two traits are not linked. (The Complex Expression Patterns of Multiple Alleles - BIOL110F2012 - Confluence, 2012) Another example is eye color. The trait is influenced by more than one gene. Some of these genes are OCA2 and HERC2.
- Epistasis is when one gene has an effect on the expression of another gene. This occurs when genes interact to produce a particular trait. An example would be the determination of coat color in certain animals (e.g. horses) wherein the effect of one gene depends on the influence of another gene controlling the deposition of hair pigment.

14

PHYSICAL BASIS OF HEREDITY (CHROMOSOME THEORY OF HEREDITY)

When Gregor Mendel formulated his laws of heredity, he postulated a particulate nature for the units of inheritance. What exactly these particles were he did not know. Today scientists understand not only the physical location of hereditary units (i.e., the genes) but their molecular composition as well. The unraveling of the physical basis of heredity makes up one of the most fascinating chapters in the history of biology.

CHROMOSOMES AND GENES

As has been discussed, each individual in a sexually reproducing species inherits two alleles for each gene, one from each parent. Furthermore, when such an individual forms sex cells, each of the resultant gametes receives one member of each allelic pair. The formation of gametes occurs through a process of cell division called meiosis. When gametes unite in fertilization, the double dose of hereditary material is restored, and a new individual is created. This individual, consisting at first of only one cell, grows via mitosis, a process of repeated cell divisions. Mitosis differs from meiosis in that each daughter cell receives a full copy of all the hereditary material found in the parent cell.

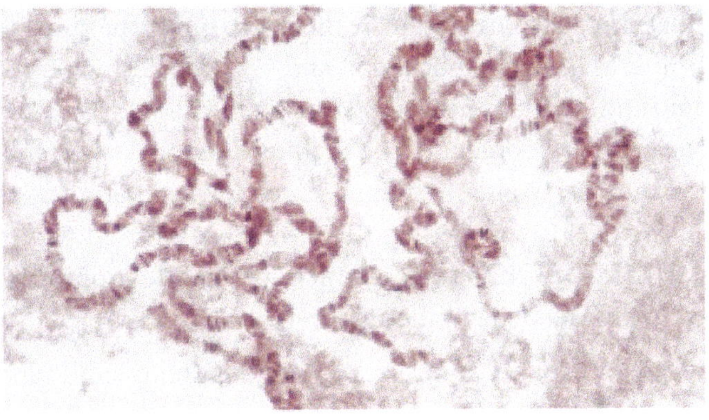

Fig. 1 Chromosomes and genes

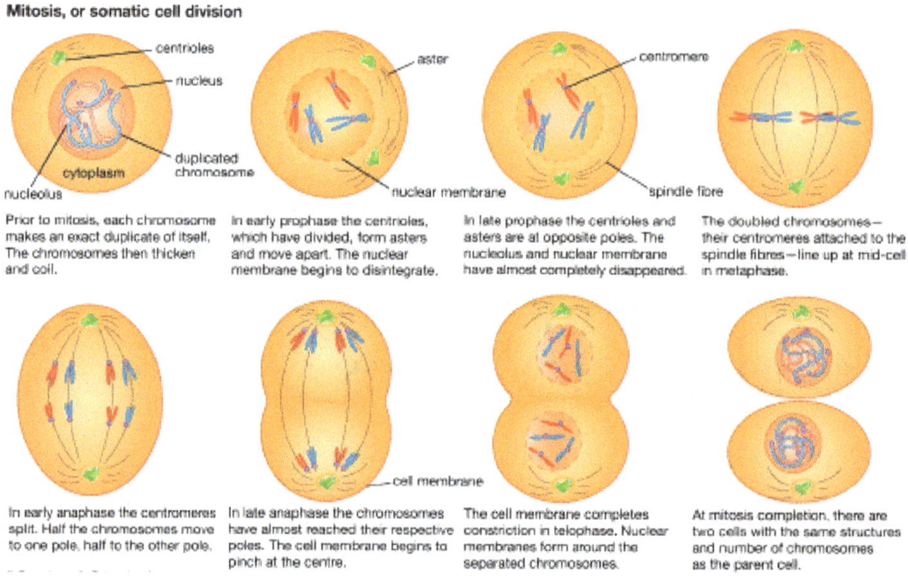

Fig. 2 One cell gives rise to two genetically identical daughter cells during the process of mitosis.

It is apparent that the genes must physically reside in cellular structures that meet two criteria. First, these structures must be replicated and passed on to each generation of daughter cells during mitosis. Second, they must be organized into homologous pairs, one member of which is parceled out to each gamete formed during meiosis.

As early as 1848, biologists had observed that cell nuclei resolve themselves into small rodlike bodies during mitosis; later these structures were found to absorb certain dyes and so came to be called chromosomes (coloured bodies). During the early years of the 20th century, cellular studies using ordinary light microscopes clarified the behaviour of chromosomes during mitosis and meiosis, which led to the conclusion that chromosomes are the carriers of genes.

CHROMOSOMAL THEORY OF INHERITANCE

The speculation that chromosomes might be the key to understanding heredity led several scientists to examine Mendel's publications and re-evaluate his model in terms of the behavior of chromosomes during mitosis and meiosis. In 1902, Theodor Boveri observed that proper embryonic development of sea urchins does not occur unless chromosomes are present. That same year, Walter Sutton observed the separation of chromosomes into daughter cells during meiosis. Together, these observations led to the development of the Chromosomal Theory of Inheritance, which identified chromosomes as the genetic material responsible for Mendelian inheritance.

The Chromosomal Theory of Inheritance was consistent with Mendel's laws and was supported by the following observations:

Fig. 3 (a) Walter Sutton and (b) Theodor Boveri are credited with developing the Chromosomal Theory of Inheritance, which states that chromosomes carry the unit of heredity (genes).

- During meiosis, homologous chromosome pairs migrate as discrete structures that are independent of other chromosome pairs.
- The sorting of chromosomes from each homologous pair into pre-gametes appears to be random.
- Each parent synthesizes gametes that contain only half of their chromosomal complement.
- Even though male and female gametes (sperm and egg) differ in size and morphology, they have the same number of chromosomes, suggesting equal genetic contributions from each parent.
- The gametic chromosomes combine during fertilization to produce offspring with the same chromosome number as their parents.

Despite compelling correlations between the behavior of chromosomes during meiosis and Mendel's abstract laws, the Chromosomal Theory of Inheritance was proposed long before there was any direct evidence that traits were carried on chromosomes. Critics pointed out that individuals had far more independently segregating traits than they had chromosomes. It was only after several years of carrying out crosses with the fruit fly, *Drosophila melanogaster*, that Thomas Hunt Morgan provided experimental evidence to support the Chromosomal Theory of Inheritance.

GENETIC LINKAGE AND DISTANCES

Mendel's work suggested that traits are inherited independently of each other. Morgan identified a 1:1 correspondence between a segregating trait and the X chromosome, suggesting that the random segregation of chromosomes was the physical basis of Mendel's model. This also demonstrated that linked genes disrupt Mendel's predicted outcomes. The fact that each chromosome can carry many linked genes explains how individuals can have many more traits than they have chromosomes. However, observations by researchers in Morgan's laboratory suggested that alleles positioned on the same chromosome were not always inherited together. During meiosis, linked genes somehow became unlinked.

HOMOLOGOUS RECOMBINATION

In 1909, Frans Janssen observed chiasmata—the point at which chromatids are in contact with each other and may exchange segments—prior to the first division of meiosis. He suggested that alleles become unlinked and chromosomes physically exchange segments. As chromosomes condensed and paired with their homologs, they appeared to interact at distinct points. Janssen suggested that these points corresponded to regions in which chromosome segments were exchanged. It is now known that the pairing and interaction between homologous chromosomes, known as synapsis, does more than simply organize the homologs for migration to separate daughter cells. When synapsed, homologous chromosomes undergo reciprocal physical exchanges at their arms in a process called homologous recombination, or more simply, "crossing over."

To better understand the type of experimental results that researchers were obtaining at this time, consider a heterozygous individual that inherited dominant maternal alleles for two genes on the same chromosome (such as *AB*) and two recessive paternal alleles for those same genes (such as *ab*). If the genes are linked, one would expect this individual to produce gametes that are either *AB* or *ab* with a 1:1 ratio. If the genes are unlinked, the individual should produce *AB*, *Ab*, *aB*, and *ab* gametes with equal frequencies, according to the Mendelian concept of independent assortment. Because they correspond to new allele combinations, the genotypes Ab and aB are nonparental types that result from homologous recombination during meiosis.

Parental types are progeny that exhibit the same allelic combination as their parents. Morgan and his colleagues, however, found that when such heterozygous individuals were test crossed to a homozygous recessive parent (*AaBb* × *aabb*), both parental and nonparental cases occurred. For example, 950 offspring might be recovered that were either *AaBb* or *aabb*, but 50 offspring would also be obtained that were either *Aabb* or *aaBb*. These results suggested that linkage occurred most often, but a significant minority of offspring were the products of recombination.

Inheritance patterns of unlinked and linked genes are shown. In (a), two genes are located on different chromosomes so independent assortment occurs during meiosis. The offspring have an equal chance of being the parental type (inheriting the same combination of traits as the parents) or a nonparental type (inheriting a different combination of traits than the parents). In (b), two genes are very close together on the same chromosome so that no crossing over occurs between them. The genes are therefore always inherited together and all of the offspring are the parental type. In (c), two genes are far apart on the chromosome such that crossing over occurs during every meiotic event. The recombination frequency will be the same as if the genes were on separate chromosomes. (d) The actual recombination frequency of fruit fly wing length and body color that Thomas Morgan observed in 1912 was 17 percent. A crossover frequency between 0 percent and 50 percent indicates that the genes are on the same chromosome and crossover occurs some of the time.

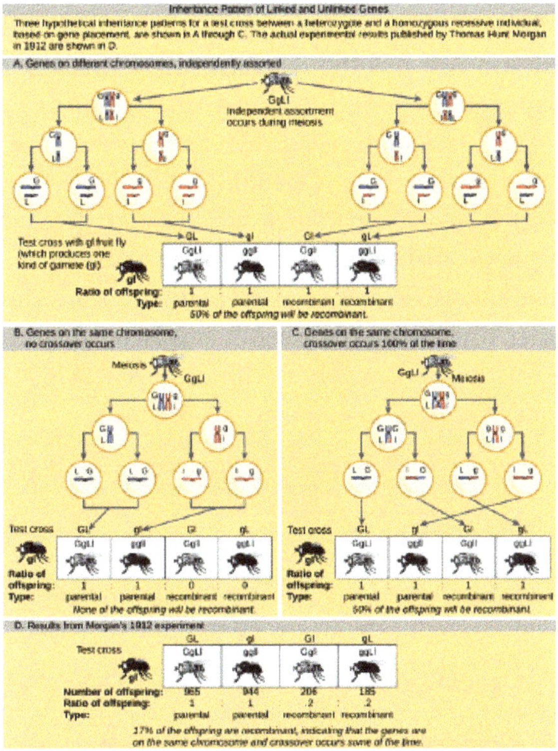

Fig. 4

In a test cross for two characteristics such as the one shown here, can the predicted frequency of recombinant offspring be 60 percent? Why or why not?

GENETIC MAPS

Janssen did not have the technology to demonstrate crossing over so it remained an abstract idea that was not widely accepted. Scientists thought chiasmata were a variation on synapsis and could not understand how chromosomes could break and rejoin. Yet, the data were clear that linkage did not always occur. Ultimately, it took a young undergraduate student and an "all-nighter" to mathematically elucidate the problem of linkage and recombination.

In 1913, Alfred Sturtevant, a student in Morgan's laboratory, gathered results from researchers in the laboratory, and took them home one night to mull them over. By the next morning, he had created the first "chromosome map," a linear representation of gene order and relative distance on a chromosome (Fig. 5).

As shown in Fig. 5, by using recombination frequency to predict genetic distance, the relative order of genes on chromosome 2 could be inferred. The values shown represent map distances in centimorgans (cM), which correspond to recombination frequencies (in percent). Therefore, the genes for body color and wing size were 65.5 " 48.5 = 17 cM apart, indicating that the maternal and paternal alleles for these genes recombine in 17 percent of offspring, on average.

To construct a chromosome map, Sturtevant assumed that genes were ordered serially on threadlike chromosomes. He also assumed that the incidence of recombination between two homologous chromosomes could occur with equal likelihood anywhere along the length of the chromosome. Operating under these assumptions, Sturtevant postulated that alleles that were far apart on a chromosome were more likely to dissociate during meiosis simply because there was a larger region over which recombination could occur. Conversely, alleles that were close to each other on the chromosome were likely to be inherited together. The average number of crossovers between two alleles—that is, their recombination frequency—correlated with their genetic distance from each other, relative to the locations of other genes on that chromosome.

Genetic Map Based on Recombination Frequencies in *Drosophila*

MUTANT		WILD TYPE
Short aristae	0	Long aristae
Black body	48.5	Gray body
Cinnabar eyes	57.5	Red eyes
Vestigial wings	65.5	Normal wings
Brown eyes	104.5	Red eyes

Values in centimorgan (cM) map units; recombination frequency of 0.01 = 1 cM

Fig. 5 This genetic map orders Drosophila genes on the basis of recombination frequency.

Considering the example cross between *AaBb* and *aabb* above, the frequency of recombination could be calculated as 50/1000 = 0.05. That is, the likelihood of a crossover between genes *A*/*a* and *B*/*b* was 0.05, or 5 percent. Such a result would indicate that the genes were definitively linked, but that they were far enough apart for crossovers to occasionally occur. Sturtevant divided his genetic map into map units, or centimorgans (cM), in which a recombination frequency of 0.01 corresponds to 1 cM.

By representing alleles in a linear map, Sturtevant suggested that genes can range from being perfectly linked (recombination frequency = 0) to being perfectly unlinked (recombination frequency = 0.5) when genes are on different chromosomes or genes are separated very far apart on the same chromosome. Perfectly unlinked genes correspond to the frequencies predicted by Mendel to assort independently in a dihybrid cross. A recombination frequency of 0.5 indicates that 50 percent of offspring are recombinants and the other 50 percent are parental types. That is, every type of allele combination is represented with equal frequency. This representation allowed Sturtevant to additively calculate distances between several genes on the same chromosome. However, as the genetic distances approached 0.50, his predictions became less accurate because it was not clear whether the genes were very far apart on the same chromosome or on different chromosomes.

In 1931, Barbara McClintock and Harriet Creighton demonstrated the crossover of homologous chromosomes in corn plants. Weeks later, homologous recombination in *Drosophila* was demonstrated microscopically by Curt Stern. Stern observed several X-linked phenotypes that were associated with a structurally unusual and dissimilar X chromosome pair in which one X was missing a small terminal segment, and the other X was fused to a piece of the Y chromosome.

By crossing flies, observing their offspring, and then visualizing the offspring's chromosomes, Stern demonstrated that every time the offspring allele combination deviated from either of the parental combinations, there was a corresponding exchange of an X chromosome segment. Using mutant flies with structurally distinct X chromosomes was the key to observing the products of recombination because DNA sequencing and other molecular tools were not yet available. It is now known that homologous chromosomes regularly exchange segments in meiosis by reciprocally breaking and rejoining their DNA at precise locations.

CHROMOSOMAL BASIS OF INHERITED DISORDERS

Inherited disorders can arise when chromosomes behave abnormally during meiosis. Chromosome disorders can be divided into two categories: abnormalities in chromosome number and chromosomal structural rearrangements. Because even small segments of chromosomes can span many genes, chromosomal disorders are characteristically dramatic and often fatal.

Identification of Chromosomes

The isolation and microscopic observation of chromosomes forms the basis of cytogenetics and is the primary method by which clinicians detect chromosomal abnormalities in humans. A karyotype is the number and appearance of chromosomes, and includes their length, banding pattern, and centromere position. To obtain a view of an individual's karyotype, cytologists photograph the chromosomes and then cut and paste each chromosome into a chart, or karyogram, also known as an ideogram (Fig. 6).

In a given species, chromosomes can be identified by their number, size, centromere position, and banding pattern. In a human karyotype, autosomes or "body chromosomes" (all of the non-sex chromosomes) are generally organized in approximate order of size from largest (chromosome 1) to smallest (chromosome 22). The X and Y chromosomes are not autosomes. However, chromosome 21 is actually shorter than chromosome 22. This was discovered after the naming of Down syndrome as trisomy 21, reflecting how this disease results from possessing one extra chromosome 21 (three total). Not wanting to change the name of this important disease, chromosome 21 retained its numbering, despite describing the shortest set of chromosomes. The chromosome "arms" projecting from either end of the centromere may be designated as short or long, depending on their relative lengths. The short arm is abbreviated *p* (for "petite"), whereas the long arm is abbreviated *q* (because it follows "p" alphabetically). Each arm is further subdivided and denoted by a number. Using this naming system, locations on chromosomes can be described consistently in the scientific literature.

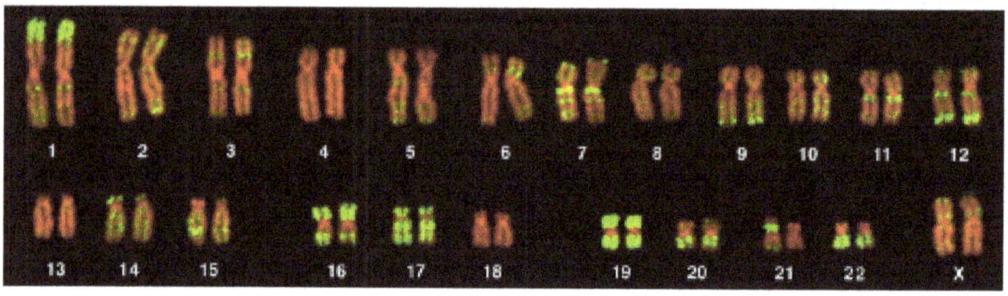

Fig. 6. This karyotype is of a female human. Notice that homologous chromosomes are the same size, and have the same centromere positions and banding patterns. A human male would have an XY chromosome pair instead of the XX pair shown.

DISORDERS IN CHROMOSOME NUMBER

Of all of the chromosomal disorders, abnormalities in chromosome number are the most obviously identifiable from a karyogram. Disorders of chromosome number include the duplication or loss of entire chromosomes, as well as changes in the number of complete sets of chromosomes. They are caused by nondisjunction, which occurs when pairs of homologous chromosomes or sister chromatids fail to separate during meiosis. Misaligned or incomplete synapsis, or a dysfunction of the spindle apparatus that facilitates chromosome migration, can cause nondisjunction. The risk of nondisjunction occurring increases with the age of the parents.

Nondisjunction can occur during either meiosis I or II, with differing results (Fig. 7). If homologous chromosomes fail to separate during meiosis I, the result is two gametes that lack that particular chromosome and two gametes with two copies of the chromosome. If sister chromatids fail to separate during meiosis II, the result is one gamete that lacks that chromosome, two normal gametes with one copy of the chromosome, and one gamete with two copies of the chromosome.

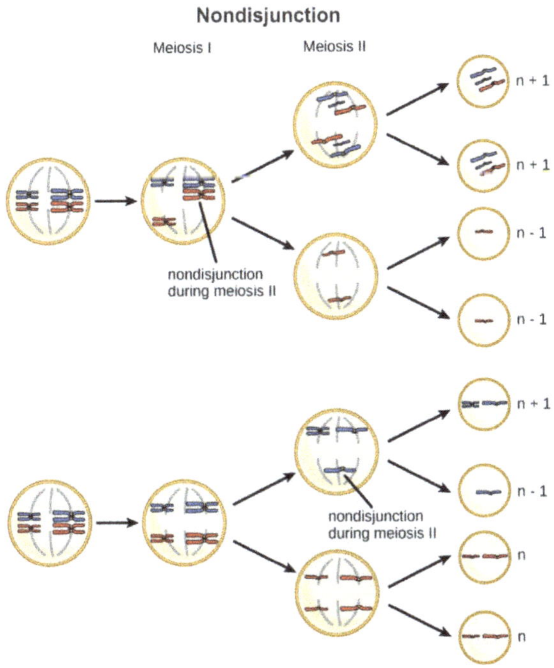

Fig. 7. Nondisjunction occurs when homologous chromosomes or sister chromatids fail to separate during meiosis, resulting in an abnormal chromosome number. Nondisjunction may occur during meiosis I or meiosis II.

ANEUPLOIDY

An individual with the appropriate number of chromosomes for their species is called euploid; in humans, euploidy corresponds to 22 pairs of autosomes and one pair of sex chromosomes. An individual with an error in chromosome number is described as aneuploid, a term that includes monosomy (loss of one chromosome) or trisomy (gain of an extraneous chromosome). Monosomic human zygotes missing any one copy of an autosome invariably fail to develop to birth because they lack essential genes. This underscores the importance of "gene dosage" in humans. Most autosomal trisomies also fail to develop to birth; however, duplications of some of the smaller chromosomes (13, 15, 18, 21, or 22) can result in offspring that survive for several weeks to many years. Trisomic individuals suffer from a different type of genetic imbalance: an excess in gene dose. Individuals with an extra chromosome may synthesize an abundance of the gene products encoded by that chromosome. This extra dose (150 percent) of specific genes can lead to a number of functional challenges and often precludes development. The most common trisomy among viable births is that of chromosome 21, which corresponds to Down Syndrome. Individuals with this inherited disorder are characterized by short stature and stunted digits, facial distinctions that include a broad skull and large tongue, and significant developmental delays. The incidence of Down syndrome is correlated with maternal age; older women are more likely to become pregnant with fetuses carrying the trisomy 21 genotype

POLYPLOIDY

An individual with more than the correct number of chromosome sets (two for diploid species) is called polyploid. For instance, fertilization of an abnormal diploid egg with a normal haploid sperm would yield a triploid zygote. Polyploid animals are extremely rare, with only a few examples among the flatworms, crustaceans, amphibians, fish, and lizards. Polyploid animals are sterile because meiosis cannot proceed normally and instead produces mostly aneuploid daughter cells that cannot yield viable zygotes. Rarely, polyploid animals can reproduce asexually by haplodiploidy, in which an unfertilized egg divides mitotically to produce offspring. In contrast, polyploidy is very common in the plant kingdom, and polyploid plants tend to be larger and more robust than euploids of their species (Fig 8).

SEX CHROMOSOME NONDISJUNCTION IN HUMANS

Fig. 8. As with many polyploid plants, this triploid orange daylily (Hemerocallisfulva) is particularly large and robust, and grows flowers with triple the number of petals of its diploid counterparts

Fig. 9. In cats, the gene for coat color is located on the X chromosome. In the embryonic development of female cats, one of the two X chromosomes is randomly inactivated in each cell, resulting in a tortoiseshell pattern if the cat has two different alleles for coat color. Male cats, having only one X chromosome, never exhibit a tortoiseshell coat color.

Humans display dramatic deleterious effects with autosomal trisomies and monosomies. Therefore, it may seem counterintuitive that human females and males can function normally, despite carrying different numbers of the X chromosome. Rather than a gain or loss of autosomes, variations in the number of sex chromosomes are associated with relatively mild effects. In part, this occurs because of a molecular process called X inactivation. Early in development, when female mammalian embryos consist of just a few thousand cells (relative to trillions in the newborn), one X chromosome in each cell inactivates by tightly condensing into a quiescent (dormant) structure called a Barr body.

The chance that an X chromosome (maternally or paternally derived) is inactivated in each cell is random, but once the inactivation occurs, all cells derived from that one will have the same inactive X chromosome or Barr body. By this process, females compensate for their double genetic dose of X chromosome. In so-called "tortoiseshell" cats, embryonic X inactivation is observed as color variegation (Fig. 9). Females that are heterozygous for an X-linked coat color gene will express one of two different coat colors over different regions of their body, corresponding to whichever X chromosome is inactivated in the embryonic cell progenitor of that region.

DUPLICATIONS AND DELETIONS

In addition to the loss or gain of an entire chromosome, a chromosomal segment may be duplicated or lost. Duplications and deletions often produce offspring that survive but exhibit physical and mental abnormalities. Duplicated chromosomal segments may fuse to existing chromosomes or may be free in the nucleus. Cri-du-chat (from the French for "cry of the cat") is a syndrome associated with nervous system abnormalities and identifiable physical features that result from a deletion of most of 5p (the small arm of chromosome 5) (Figure 10). Infants with this genotype emit a characteristic high-pitched cry on which the disorder's name is based.

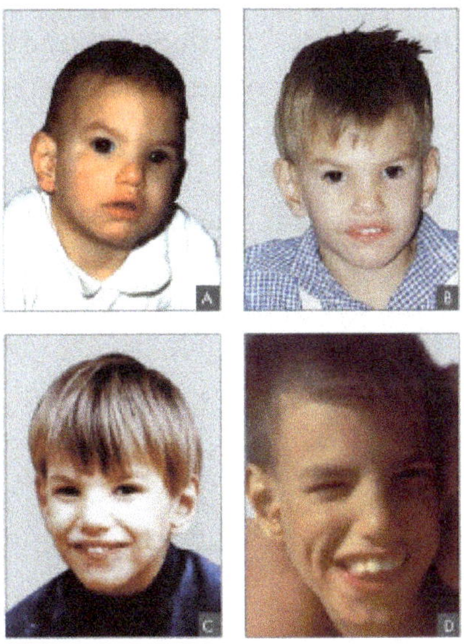

Fig. 10. This individual with cri-du-chat syndrome is shown at two, four, nine, and 12 years of age

CHROMOSOME INVERSIONS

A chromosome inversion is the detachment, 180° rotation, and reinsertion of part of a chromosome. Inversions may occur in nature as a result of mechanical shear, or from the action of transposable elements (special DNA sequences capable of facilitating the rearrangement of chromosome segments with the help of enzymes that cut and paste DNA sequences). Unless they disrupt a gene sequence, inversions only change the orientation of genes and are likely to have more mild effects than aneuploid errors. However, altered gene orientation can result in functional changes because regulators of gene expression could be moved out of position with respect to their targets, causing aberrant levels of gene products.

An inversion can be pericentric and include the centromere, or paracentric and occur outside of the centromere. A pericentric inversion that is asymmetric about the centromere can change the relative lengths of the chromosome arms, making these inversions easily identifiable.

When one homologous chromosome undergoes an inversion but the other does not, the individual is described as an inversion heterozygote. To maintain point-for-point synapsis during meiosis, one homolog must form a loop, and the other homolog must mold around it. Although this topology can ensure that the genes are correctly aligned, it also forces the homologs to stretch and can be associated with regions of imprecise synapsis.

TRANSLOCATIONS

A translocation occurs when a segment of a chromosome dissociates and reattaches to a different, nonhomologous chromosome. Translocations can be benign or have devastating effects depending on how the positions of genes are altered with respect to regulatory sequences. Notably, specific translocations have been associated with several cancers and with schizophrenia. Reciprocal translocations result from the exchange of chromosome segments between two nonhomologous chromosomes such that there is no gain or loss of genetic information (Figure 11).

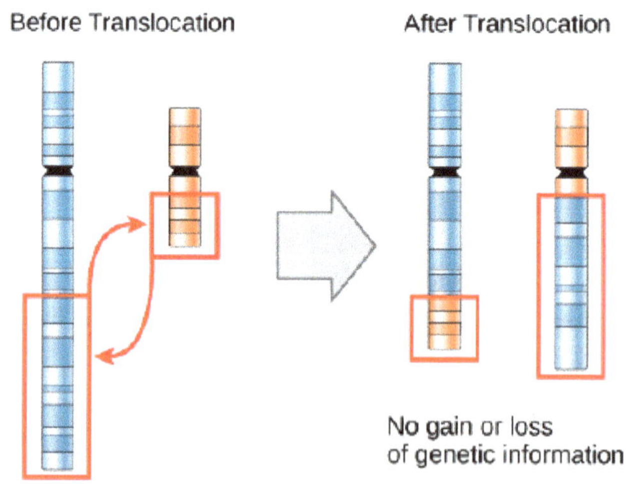

Fig. 11. A reciprocal translocation occurs when a segment of DNA is transferred from one chromosome to another, nonhomologous chromosome.

SUMMARY

The number, size, shape, and banding pattern of chromosomes make them easily identifiable in a karyogram and allows for the assessment of many chromosomal abnormalities. Disorders in chromosome number, or aneuploidies, are typically lethal to the embryo, although a few trisomic genotypes are viable. Because of X inactivation, aberrations in sex chromosomes typically have milder phenotypic effects. Aneuploidies also include instances in which segments of a chromosome are duplicated or deleted. Chromosome structures may also be rearranged, for example by inversion or translocation. Both of these aberrations can result in problematic phenotypic effects. Because they force chromosomes to assume unnatural topologies during meiosis, inversions and translocations are often associated with reduced fertility because of the likelihood of nondisjunction.

15

LINKAGE AND CROSSING OVER

INTRODUCTION

Linkage and Crossing Over are two related, yet different phenomena occurring in Eukaryotic Organisms. Genetic Linkage or, simply, Linkage is the tendency of genes to stay together in a Chromosome while Crossing Over is a phenomenon through which genetic information is exchanged in the germline. Both of these play an important role in heredity. Both the events are somewhat related to each other. However, Linkage is more about the tendency of genes in a Chromosome to inherit together. Through Crossing Over, the Genes separate and segregate into different gametes.

GENETIC LINKAGE

Genetic Linkage means the tendency of the genes or DNA sequences on a chromosome to be inherited together during meiosis of sexual reproduction. Linked genes are the ones located on the same chromosome. For Example, genes on hair colour and eye colour. That is why some individuals inherit hair and eye colours together. Such as black hair with brown eyes or individuals with brown hair and blue eyes inherited from parents.

CROSSING OVER

The exchange of chromosomes between non-sister chromatids to form gametes is Crossing Over. Crossing Over shuffles the alleles on parental chromosomes, so that the gametes carry the combination of genes from mother and father. In simple language, it is the swapping of genetic material in the germline. During meiosis, or the formation of an egg cell and sperm cells, paired chromosomes get aligned so that similar DNA sequences from the paired chromosomes come against one another. This phenomenon accounts for genetic variation in sexually reproducing organisms and it is also essential for the normal segregation of chromosomes.

LINKAGE AND CROSSING OVER

Linkage is the tendency of genes present in the chromosome to stay intact and transfer to the next generation whereas crossing over is the exchange of chromosomal sections to disrupt built links and form new linkages.

Linkages generate parental types and increases in age as the linkage increases. Crossing over generates recombinations and it decreases with age and the occurrence of crossing over between two genes is decreased if they are placed closely, whereas the linkage strength between two genes increases if a chromosome is located in the vicinity.

Linkages assist in maintaining a newly improved variety, crossing over, on the other hand, forms as a source of variations to produce new varieties.

Crossing over, in other terms, is the exchange of segments observed in homologous chromosomes between non-sister chromatids and takes place during the pachytene stage of the prophase I in the **cell division** process of meiosis and always takes place within linked genes. The recombination of linked genes that crossing over produces plays a significant role in evolution.

THE DIFFERENCE BETWEEN LINKAGE AND CROSSING OVER

- Crossing Over is the process of separation of genes between homologous pairs into various gametes. Linkage is the tendency of inheriting genes together on the same chromosome.
- Linkage occurs when two genes are closer to each other on the same chromosome. On the other hand, Crossing Over takes place when two genes are located far apart on the same chromosome.
- Crossing Over may disrupt the gene groups made by Linkage.
- Unlike the Linkage, Crossing Over only occurs during the prophase of meiosis I and it produces recombinant alleles, which is not the case in the Linkage.

Linkage	Crossing Over
Ensures that the genes stay in a chromosome to inherit together	Crossing Over enables the separation of genes in a chromosome and separates it into various gametes
The strength of the genetic Linkage between two genes is thought to be inversely proportional to the distance between them in the chromosome	The probability of two genes Crossing Over is directly proportional to the distance between them in the chromosome
The strength of the Linkage between two genes can increase if the distance between them decreases	The probability of two genes Crossing Over decreases if the distance between them decreases
Linkage	**Crossing Over**
Ensures the continuity of the parental trait in the offspring	Causes alterations in the parental traits in the offspring
Reduces the probability of variation through sexual reproduction	Increases the probability of introducing vari-ability in sexual reproduction

LINKAGE AND RECOMBINATION

"Linkage and recombination are the phenomena that describe the inheritance of genes."

In linkage, two or more genes linked together are always inherited together in the same combination for more than two generations, whereas in recombination the genetic material is exchanged between different organisms which leads to the production of offsprings with the combination of traits.

The chromosomal theory of inheritance came into existence long after Mendelian genetics. The lack of advanced technology and physical proof were two reasons for the rejection of Mendel's concept of genetics. However, later experiments confirmed Mendel's results.

Biologists developed the chromosomal theory of inheritance. The work was further carried forward and proved by T.H.Morgan. This led to the concept of linkage and recombination of the genes.

Linkage

Morgan noted that while crossing a set of characters, two genes did not follow Mendel's law as they did not divide as per the law. The likelihood of achieving a parental combination if two genes are situated on the same chromosome was relatively higher in the subsequent generation in comparison to the non-parental combination. The physical connection of genes was referred to as linkage.

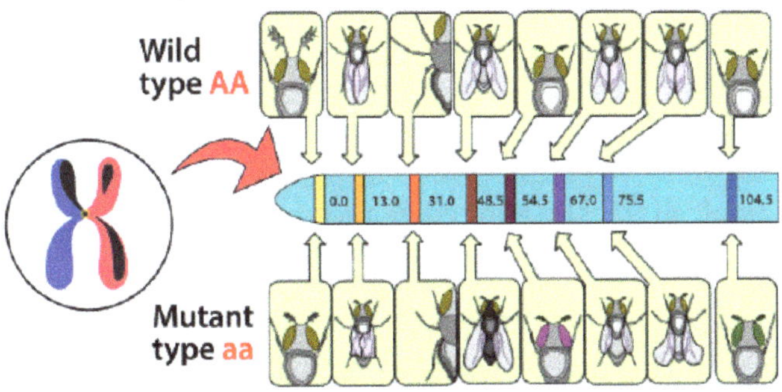

Fig. 15.1 Linkage and Recombination

Genes are said to be linked when genes for different traits are located in similar chromosomes and hence are tied to each other. It is a deviation from the Mendelian principle of independent assortment that is appropriate to be applied to the genes that are situated on different chromosomes.

The term genetic recombination described the non-parental gene combinations in a dihybrid cross.

Additionally, they observed that the likelihood of recombination is reliant on how influential the linkage is. In other terms, though two genes are linked on chromosomes, the genes may or may not be linked tightly.

A few genes have powerful linkage giving a limited chance of recombination whereas another linkage of genes is loosely linked and weak providing a greater chance of recombination.

Once the linked genes were identified, the recurrence of linked genes also determined the expression of characteristics in the next generation.

Types of Recombination

Recombination can be of the following two types:

- **Homologous Recombination:** This type of recombination occurs between chromosomes of similar sequences and is carried out during meiosis.
- **Non-homologous Recombination:** This occurs between chromosomes that are not similar.
- **Site-specific Recombination:** This is observed between very short sequences that usually contain similarities.
- **Mitotic Recombination:** Mitotic recombination occurs during interphase. However, this type of recombination is usually harmful and can result in tumors. It increases when the cells are exposed to radiation.

The **prokaryotic cells** undergo recombination through one of the three processes:

- Conjugation
- Transformation
- Transduction

Recombination of Linked Genes

Let us take an example of freckles and red hair to understand the recombination of linked genes.

Freckles and red hair happens in people because the genes for both are present on the same chromosome. During homologous recombination, the DNA is rarely split between the two genes.

Although homologous recombination occurs a number of times, the two features are inherited together most of the time because the chances of splitting up of the DNA coding for these two genes are very low. As a result, genes are inherited together most of the time.

Morgan's Experiment

Thomas Hunt Morgan worked on *Drosophila melanogaster* (fruit flies) to explain how the process of sexual reproduction produced variations. He preferred to work on fruit flies or *Drosophila melanogaster* for the subsequent reasons:

1. It could be cultivated easily on a synthetic medium in the laboratory.
2. Exhibit a short lifespan (two weeks).
3. A single mating could generate multiple progenies.
4. Differentiable male and female sexes.
5. Hereditary variations are of different types.

Morgan carried out a dihybrid cross between white-eyed and yellow-bodied females and red-eyed and brown-bodied males. Self-crossing of the F_1 generation surprisingly produced an F_2 generation without being the ratio of 9:3:3:1. The outcome exhibited a divergence from Mendel's dihybrid cross in peas.

Types of Linkages

Linkages can be classified into two types established on the absence or presence of non-parental combinations or new combinations, namely:

Complete Linkage

A linkage is said to be complete when two or multiple characteristics are inherited and normally surface in two or further generations in their parental or original combinations, they are known as complete linkage. These particular genes do not generate combinations that are non-parental. The genes that exhibit these linkages are located nearby in the same chromosome. Examples – genes for long wings and grey body in male Drosophila

Incomplete Linkage

It is displayed by genes that generate some portion of non-parental combinations. These genes are situated at a distance on the chromosomes which can be attributed to the occasional or accidental deconstruction of chromosomal segments while crossing over.

16

SEX LINKED, SEX INFLUENCED AND SEX LIMITED TRAITS

Sex is determined by genetic information in human. It extensively controls physical, psychological, and behavioral traits which are also influenced by genetic information. Sex-related genetics may be categorized by how sex affects phenotypic traits, including sex itself as a trait (i.e., sex determination), sex-linked trait, sex-limited trait, and sex-influenced trait. Sex determination depends on the existence of Y chromosome, especially the sex-determining region Y gene on the chromosome. Males with translocation of the gene can have chromosomes of XX. Females with mutation of the gene can have chromosomes of XY.

Sex defined as a binary trait has been proven to be more complex as accumulating research reveals its genetics. Different molecular mechanisms operate in sexual development not only by the gene, but also by other autosomal genes that can control hormone balances. Sex might be expressed in terms of an ordinal trait. Sex can be categorized as typical female, female with subtle variation, female with moderate variation, sex with XX and testicular disorder of sex development (DSD), sex with ovotesticular DSD, sex with XY and DSD, male with moderate variation, male with subtle variation, and typical male.

Sex-linked traits are determined by genes located on the sex chromosomes. Sex-limited traits are determined by genes located on autosomes and express only in one sex. While these traits are responsible considerably for sexual dimorphism, sex-influenced traits do not show distinctive expression between women and men. Numerous analytical models have treated sex as an essential factor. However, dissection of genetic factors associated with sex has been limitedly explored to examine sex-influenced traits.

This is because many of the traits are intricate that they can hardly be explained by a few genes or sex-specific genes (mitochondrial and sex-chromosomal genes). Also, few appropriate analytical methods have been employed to explain sex-influenced traits.

Here, I would like to discuss assumption violations in identifying genetic factors for sex-influenced traits and propose appropriate methods to overcome problems.

GENETIC ANALYSIS OF SEX-INFLUENCED TRAITS 2.

Most analytical models for sex-influenced traits assume an additive sex effect by treating it as a covariate in models or adjusting it preliminarily (Fig. 1A). This leads to inference on genetic and environmental effects associated with the term "sex" in the lump. Sex has been interchangeably used with gender contrary to their original meanings of biological and social characteristics of women and men, either of which has connotations of biology and sociology.

Sex is more likely to possess biological connotations. Gender is more likely to possess social connotations. A problem is that the analytical models most likely ignore interactions of sex with genetic factors. 'Sex' considerably depends on hormones which control the whole body. Thus, genetic effects related to hormones are likely to interact with other genetic effects. Furthermore, different social environments by 'gender' interact with genetic effects. This is the reason why gene-by-sex interaction might be explained as gene-by-gene interaction and/or gene-by-environment interaction in analytical models.

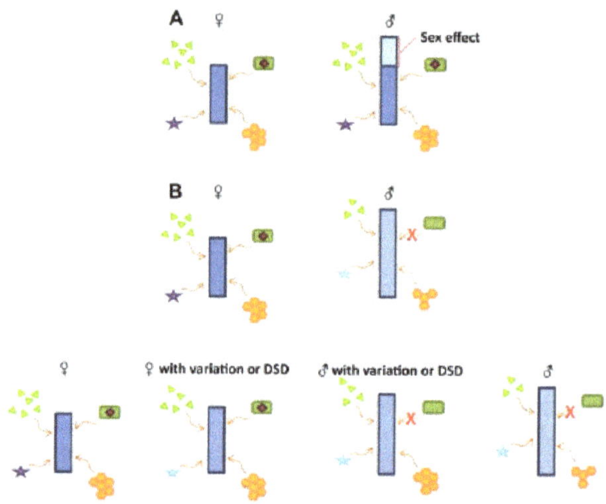

Fig. 1. Schematic concepts for analyzing quantitative trait with sex as a covariate (A) and for analyzing quantitative trait with data partitioned by sex (B) or by a wider spectrum of sex (C). Bar size indicates quantitative phenotype. Circular and polygonal figures indicate genetic products. Environmental effect is not presented to avoid confounding. DSD: disorder of sex development.

The gene-by-sex interaction has been shown in hypertension, schizophrenia, rheumatoid arthritis, and recombination rate. From an evolutionary point of view, the gene-by-sex interaction can be produced by sex-specific or sexually antagonistic selection. Nevertheless, the gene-by-sex interaction has been ignored mostly even in studies designed to identify other gene-by-gene or gene-by-environment interaction.

SEX DETERMINATION VS. SEX DIFFERENTIATION

Human sex refers to the processes by which an individual becomes either a male or female during development. Complex mechanisms are responsible for male sex determination and differentiation. The steps of formation of the testes are dependent on a series of Y-linked, X-linked and autosomal genes actions and interactions. After formation of testes the gonads secrete hormones, which are essential for the formation of the male genitalia. Hormones are transcription regulators, which function by specific receptors. Ambiguous genitalia are result of disruption of genetic interaction. This review describes the mechanisms, which lead to differentiation of male sex and ways by which the determination and differentiation may be interrupted by naturally occurring mutations, causing different syndromes and diseases.

What is Sex Determination?

Sex determination is a complex system of genes, proteins coded by them and their functions on various target organs that cumulatively determine the development of sexual characteristics in an organism.

The process of sex determination in an organism starts with the development of the male and female reproductive structures right at the stage of embryonic development. The sex determination system is switched on in an undifferentiated zygote depending on the type of sex chromosomes present in it. What follows is a cascade of events comprising protein synthesis and activation and muting of genes and synthesis of various proteins (factors) that act differently on various organs and lead to their sex-specific development.

What are Chromosomes?

Chromosomes are rod-shaped structures which become clearly visible during cell division in the cell. They are highly supercoiled DNA molecules that remain coiled around histone and non-histone proteins. The chromosomes are present as less coiled thread-like structures called chromatin in the nucleus of a non-dividing cell forming a network called chromatin reticulum. The chromosome number in a species remains constant. However, chromosomal aberrations are sometimes seen that create abnormalities in human beings.

TYPE OF SEX CHROMOSOMES IN MALES AND FEMALES

There are 46 chromosomes in the human cells. Out of these 44 chromosomes are autosomes and have genes for characters which are not related to the formation of reproductive organs. The rest of the two are called sex chromosomes. There are two types of sex chromosomes. X and Y. Females have 44 autosomes and 2 X chromosomes and males have 44 autosomes and one X and one Y chromosome. Hence all the gametes produced in females (ova) contain X chromosomes, while sperms can be of two types X containing and Y containing. When the X containing sperm fertilizes the ova, female offspring are formed. When the ovum is fertilized by Y containing sperm, male offspring are formed. Hence the sex of the child is determined by the type of sex chromosome contributed by the father.

Role of X Chromosomes

The x chromosome is made up of 155 million DNA base pairs and accounts for 5% of the DNA in a cell. The total number of genes present in the X chromosomes is nearly 800 to 900. In 1901 Clarence Erwin McClung suggested for the first time that X chromosomes were involved in sex determination. The gene BMP15 lies in the short arm of the X chromosome and helps in the development of ovaries in females. Gene ZFX has been shown to impact the number of germ cell production in other mammals with less effect in humans. Apart from these genes, the X chromosome also contains genes for bodily characters like coloured vision, the gene for anti-haemophilic factor and a number of genes involved in the development of mental abilities of a person.

Other than these, there are hundreds of genes which are not involved in sex determination in females. Human males have only one X chromosome while human females have two. Hence, in the early stages of development, one of the two X chromosomes is permanently inactivated in the somatic cells so that gene expression of one X chromosome out of the two is restricted after the successful development of the sex organs in females. This ensures that the number of proteins produced by the gene expression of X chromosomes in human males and females remains the same. This phenomenon is known as dosage compensation. The process of inactivation is called lyonization and the inactivated X chromosome in the form of facultative heterochromatin is known as the Barr body.

Role of Y Chromosomes

In 1905, Nettie Stevens identified Y chromosomes as a sex-determining chromosome while working on mealworms. She proposed the theory of C.E. McClung identifying X chromosome as the sex-determining chromosome was wrong. The Y chromosomes contain the SRY gene. SRY or the sex-determining region of Y, also known as the Testis Determining Factor (TDF) produces a protein which initiates testis development during embryogenesis in placental mammals and marsupials. The primordial gonad during embryogenesis is bipotential and can be converted into testes or ovaries. The male-specific transcription factors are activated by the TDF so that the bipotential cells of the gonad develop into Sertoli cells. The cells which have functional TDF proteins differentiate to form the cells of the testis.

SEX CHROMOSOMAL DEFECTS IN HUMAN BEINGS

Numerical abnormalities of X and Y chromosomes are responsible for causing several conditions in human beings. Some of them are discussed below.

i. **Klinefelter Syndrome:** A condition where 44+XXY karyotype gives rise to male individuals with a reproductive disability, tall structure, low IQ learning and reading disabilities.

ii. **Triple X Syndrome:** A condition where 44+XXX karyotype gives rise to female individuals who are fertile, with average IQ and stature more than that of an average female.

iii. **Turner's Syndrome:** A condition where 44+X0 karyotype gives rise to female individuals with short structure. They are infertile as they have ill-developed reproductive parts.

iv. **XYY Syndrome:** A condition where 44+XYY karyotype gives rise to male individuals with increased stature, learning problems etc.

PROCESS OF DETERMINATION OF SEX OF A FOETUS - AMNIOCENTESIS

The sex of the foetus can be determined before birth by the process called Amniocentesis. The fetus floats in the amniotic sac filled with amniotic fluid in the uterus. The cells from the fetus shed off and float in the amniotic fluid.

A needle is used to draw a small volume of this fluid guided by ultrasound. Karyotyping of the cells of this fluid helps in the determination of the presence of X and Y chromosomes in the cells of the fetus. This process is however banned in India considering the increased cases of female foeticides in different regions.

Thus sex determination before birth is a punishable offence and can lead to imprisonment up to seven years and a fine of up to 5 lakhs.

CHROMOSOMAL MECHANISM OF SEX DETERMINATION

As we know, once the egg of a female gets fertilized by the sperm of a male, it results in the formation of zygote. This zygote is the first cell of a new life and this cell contains the content of both the male and female cells. And this zygote will divide again and again until it eventually forms a baby.

In sexually reproducing organisms besides morphological and behavioral differences between male and female sexes, the chromosomal difference also occurs.

A German biologist, Henking (1891) firstly noted that half of the spermatozoa contained an extra chromosome which was called by him is X body. The significance of X body was not known till 1902. In 1902 Mcclung suggested that the X body was involved in some way in determination of sex. Later on it was well known that the X body was a chromosome involved in sex determination so it is called sex chromosome.

In human females all chromosomes are paired and are in equal size. In the mail all the chromosomes are paired but the chromosome prepared with X chromosome was distinctly smaller and was called Y chromosome.

In dioecious diploid organisms following two systems of chromosomal determination of sex have been recognised:

- Heterogametic males
- Heterogametic females

Heterogametic Males : in this type, the female sex has two X chromosomes, while the male has only one X chromosome. Therefore, during the gametogenesis male produces two types of gametes, 50% with X chromosome while 50% without X chromosome.

So a sex which produces two different type of gametes is called heterogametic sex, while sex which produces similar type of gametes is called homogametic sex. The heterogametic male are of following two types:

- **XX-XO Type:** in certain insects such as Hemiptera and Orthoptera, the female has two X chromosomes while male has only one X chromosome.
- **XX-XY Type:** In mammals, certain insects and some angiosperm plants, the female has two X chromosomes, while male has one X chromosome and one Y chromosome.

Heterogametic Females : in this type, male sex has two sex chromosomes (X chromosomes), while the female has only one X- chromosome. Therefore, during gametogenesis females produce two types of eggs/ova, 50% with X chromosome while 50% without X chromosome.

- **ZO-ZZ Type:** in birds, moths and butterflies, the female has one Z chromosome while male has two Z chromosomes.
- **ZW-ZZ Type:** In fishes, reptiles, some birds and insects, the female has one Z chromosome and one W chromosome, while male have two Z chromosomes.

ENVIRONMENTAL SEX DETERMINATION IN BONELLIA

In some organisms the environment influences the determination of sex,such as temperature, population size of species, sex of other individuals etc.

In Marine worm bonellia, Maine bonellia lives inside the cloaca of females like a parasite. The female produces those eggs which are being fertilized by the male. The eggs hatch into the Larva. Dost Lava settles on the proboscis of female, develops into male. If the larva does not find any female to settle, it attaches to the substratum and develops into a female.

Sex determination: Initial event that determines whether the gonads will develop as testes or ovaries. *Sex is determined by "the heat of the male partner during intercourse" -Aristotle (335 B.C.).* Today: both environmental and internal mechanisms of sex determination can operate in different species.

There are two different types of sexes, that participate in sexual reproduction. So it is natural to find it confounding as from which sex the baby inherits, that results in the sex of the child. There are several other procedures followed to determine the sex of a newborn baby. Based on the environmental signals the sex of a baby can be determined. In a few animal species, temperature plays a major role in sex determination. In other animals, like snails, it is possible to change sex as they are not genetically processed. In human beings, the sex of an individual is genetically determined. In other words, the genes which are inherited from their parents decide the sex of the child.

DETERMINING THE SEX

To understand how the determination of sex happens, we need to know the following process.

1. Humans have 23 pairs of chromosomes. Out of these 23 pairs, 22 pairs are Autosomes and only one pair is the 'Sex Chromosome', which actively takes part in the process of sex determination.
2. Both males and females carry two sets of sex chromosome.
3. 3. Male has one X and one Y (XY) sex chromosome in which both are active
4. Female has both X (XX) sex chromosome in which one is active.
5. The XY sex-determination system is found in humans, mammals, in some insects, and in few plant species.
6. All children will inherit an X chromosome from their mother, despite whether they are a boy or girl. Thus, the sex of the children will be determined by the type of chromosome inherited from their father.
7. A child who inherits Y chromosome will be a boy and who inherits X chromosome will be a girl.
8. A male semen's holds approximately 50 percent of Y sperm cells and other 50 percent of X sperm cells. Therefore, there are possibilities that men have three or more children of a single-gender.

Sex differentiation

Subsequent events that ultimately produce either the male or female sexual phenotype. Sexual differentiation is conformed in the human during four successive steps: the constitution of the genetic sex, the differentiation of the gonads, the differentiation of the internal and the external genital tractus and the differentiation of the brain and the hypothalamus.

Sex determination, which depends on the sex-chromosome complement of the embryo, is established by multiple molecular events that direct the development of germ cells, their migration to the urogenital ridge, and the formation of either a testis, in the presence of the Y chromosome (46, XY), or an ovary in the absence of the Y chromosome and the presence of a second X chromosome (46, XX). Sex determination sets the stage for sex differentiation, the sex-specific response of tissues to hormones produced by the gonads after they have differentiated in a male or female pattern. A number of genes have been discovered that contribute both early and late to the process of sex determination and differentiation. In many cases our knowledge has derived from studies of either spontaneous or engineered mouse mutations that cause phenotypes similar to those in humans. How mutations in these genes cause important clinical syndromes and the clinical entities that continue to elude classification at the molecular level have to be tested. Knowledge of the molecular basis of disorders of sex determination and differentiation pathways will continue to have a strong influence on the diagnosis and management of these conditions.

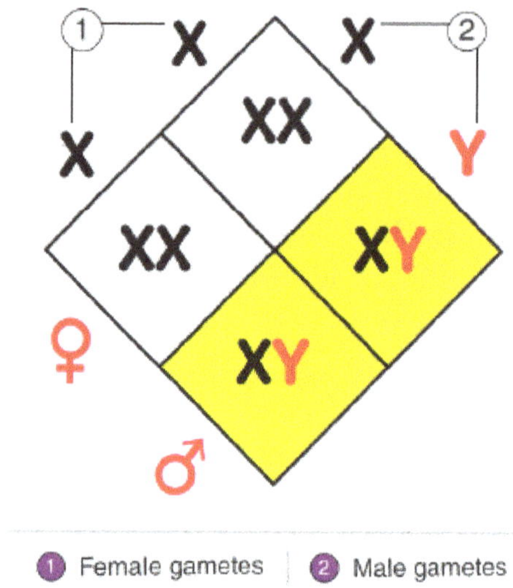

Fig. 2 Sex determination - Boy or Girl

DOSAGE COMPENSATION AND GENETIC IMPRINTING

Dosage compensation is the process by which organisms equalize the expression of genes between members of different biological sexes. Across species, different sexes are often characterized by different types and numbers of sex chromosomes. In order to neutralize the large difference in gene dosage produced by differing numbers of sex chromosomes among the sexes, various evolutionary branches have acquired various methods to equalize gene expression among the sexes. Because sex chromosomes contain different numbers of genes, different species of organisms have developed different mechanisms to cope with this inequality. Replicating the actual *gene* is impossible; thus organisms instead equalize the *expression* from each gene. For example, in humans, females (XX) silence the transcription of one X chromosome of each pair, and transcribe all information from the other, expressed X chromosome. Thus, human females have the same number of expressed X-linked genes as do human males (XY), both sexes having essentially one X chromosome per cell, from which to transcribe and express genes.

Different lineages have evolved different mechanisms to cope with the differences in gene copy numbers between the sexes that are observed on sex chromosomes. Some lineages have evolved dosage compensation, an epigenetic mechanism which restores expression of X or Z specific genes in the heterogametic sex to the same levels observed in the ancestor prior to the evolution of the sex chromosome. Other lineages equalize the expression of the X- or Z- specific genes between the sexes, but not to the ancestral levels, i.e. they possess incomplete compensation with "dosage balance".

One example of this is X-inactivation which occurs in humans. The third documented type of gene dose regulatory mechanism is incomplete compensation without balance (sometimes referred to as incomplete or partial dosage compensation). In this system gene expression of sex-specific loci is reduced in the heterogametic sex i.e. the females in ZZ/ZW systems and males in XX/XY systems.

There are three main mechanisms of achieving dosage compensation which are widely documented in the literature and which are common to most species. These include random inactivation of one female X chromosome, a two-fold increase in the transcription of a single male X chromosome, and decreased transcription by half in both of the X chromosomes of a hermaphroditic organism.

These mechanisms have been widely studied and manipulated in model organisms commonly used in the laboratory research setting. A summary of these forms of dosage compensation is illustrated in Fig 3. However, there are also other less common forms of dosage compensation, which are not as widely researched and are sometimes specific to only one species (as observed in certain bird and monotreme species).

X-CHROMOSOME DOSAGE COMPENSATION

Introduction

In numerous organisms, sex is determined by a chromosome counting mechanism that distinguishes one sex chromosome from two. In flies and worms, XX embryos become females (or hermaphrodites), while XO or XY embryos become males. Sex can also be specified by the presence of a particular sex chromosome, such as the Y chromosome of mammals: XY embryos are males, and XX embryos females. These sex-determining mechanisms cause the two sexes to differ in their dose of X chromosomes, yet both sexes require equivalent levels of X-chromosome gene products. A chromosome-wide regulatory process called dosage compensation neutralizes the difference in X-linked gene dose between sexes by equalizing X-chromosome transcript levels.

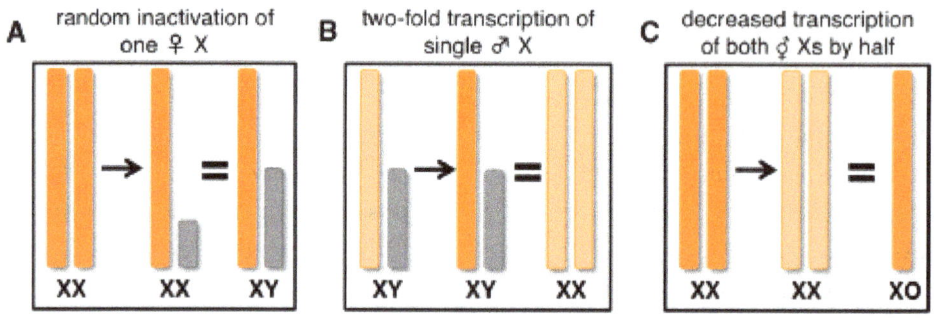

(A) Mus musculus (as well as humans and many mammalian organisms) follow a pattern of dosage compensation in which females randomly inactive one X chromosome in every cell in their body to match male expression of a single X chromosome.

(B) D. melanogaster follow a mechanism of dosage compensation in which males increase the transcription of their single X chromosome by two-fold in order to match female expression of two X chromosomes.

(C) C. elegans hermaphroditic worms express both X chromosomes by half to match male worms with only one X chromosome.

Fig. 3. Three main mechanisms of dosage compensation observed in common model eukaryotic organisms.

The strategies for dosage compensation are diverse in different organisms (Fig. 4), but in all known cases, specialized dosage compensation complexes are targeted exclusively to the X chromosome(s) of one sex to modulate gene expression in only that sex. This selective recruitment of the dosage compensation machinery establishes the epigenetic regulation of X chromosomes that is maintained throughout the lifetime of the animal. Female mammals randomly inactivate one X chromosome. Male flies double the transcription rate of their single X chromosome. Hermaphrodite worms keep both X chromosomes active, but repress transcript levels from each X chromosome by half. In flies and mammals, this gene regulation utilizes non-coding RNAs that coat the regulated X, and the chromosome-wide regulation is accompanied by the sex-specific modification of histones on the dosage compensated X chromosomes. In all three species, dosage compensation is essential, and failure to accomplish this global regulation causes either male- or female-specific lethality.

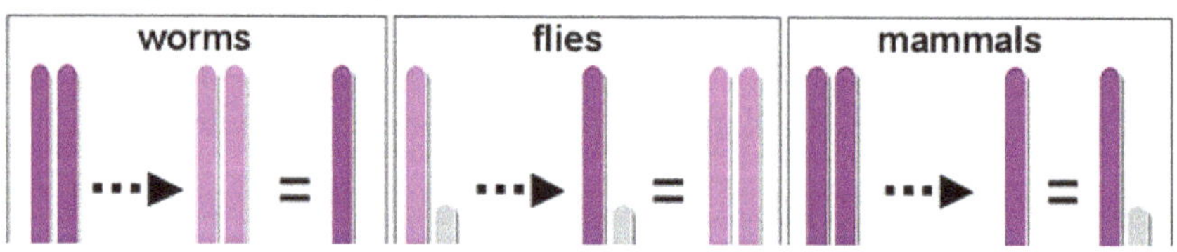

Fig. 4. Diverse strategies for dosage compensation. Organisms use different strategies to equalize X-linked gene expression between males (XY or XO) and females (or hermaphrodites; XX). Female mammals randomly inactivate one X chromosome. Male fruit flies double the transcription rate of their single X chromosome. Hermaphrodite worms half the expression of both X chromosomes.

Fundamental questions are relevant to all forms of dosage compensation. First, what is the composition of the machinery that implements dosage compensation? Second, what are the sex-specific factors that activate the dosage compensation machinery in only one sex? Third, what are the *cis*-acting recruitment sites that target X chromosomes for regulation by the dosage compensation complex? Fourth, how is gene expression coordinately controlled along an entire X chromosome? Fifth, what is the molecular mechanism for fine tuning X-linked gene expression by only two-fold? These basic questions have been addressed in *C. elegans*, using integrated genetic, biochemical, and cell biological approaches to dissect this regulatory process.

GENETIC IMPRINTING

People inherit two copies of their genes—one from their mother and one from their father. Usually both copies of each gene are active, or "turned on," in cells. In some cases, however, only one of the two copies is normally turned on. Which copy is active depends on the parent of origin: some genes are normally active only when they are inherited from a person's father; others are active only when inherited from a person's mother. This phenomenon is known as genomic imprinting.

In genes that undergo genomic imprinting, the parent of origin is often marked, or "stamped," on the gene during the formation of egg and sperm cells. This stamping process, called methylation, is a chemical reaction that attaches small molecules called methyl groups to certain segments of DNA. These molecules identify which copy of a gene was inherited from the mother and which was inherited from the father. The addition and removal of methyl groups can be used to control the activity of genes.

Only a small percentage of all human genes undergo genomic imprinting. Researchers are not yet certain why some genes are imprinted and others are not. They do know that imprinted genes tend to cluster together in the same regions of chromosomes. Two major clusters of imprinted genes have been identified in humans, one on the short (p) arm of chromosome 11 (at position 11p15) and another on the long (q) arm of chromosome 15 (in the region 15q11 to 15q13).

IMPRINTING MECHANISMS

Imprinting can occur when one of the gene's parental alleles is silenced throughout the embryonic development of the individual by an alteration in parental DNA made during parental gametogenesis (the formation of gametes, or sperm in males and eggs in females). The other parental allele is therefore allowed expression during embryonic development. A mechanism by which this occurs is DNA methylation (the addition of a CH_3, or methyl, group to specific regions of DNA) at imprinting control regions (ICRs). Intracellular DNA-reading mechanisms exist after fertilization to check that the correct parental allele has been allowed differential expression.

Imprinting and fetal development

Imprinting has been able to explain certain predicaments of life in utero. A number of imprinted genes are related to embryonic and fetal growth and thus the extraction of resources from the uterine environment for growth. Mother and father, however, have different interests in how resources are extracted, because of asymmetrical parental investment in each given child. This arises from the fact that mothers can only have one child every nine months for approximately 20 years, whereas a father could conceivably impregnate many different women from puberty until death.

Systematic knockout (inactivation) studies of key imprinted genes, especially as performed on mice, have provided support for the hypothesis that imprinted genes that allow expression of paternally inherited alleles tend to drive more extraction of nutrients from the mother during gestation and after birth to produce a larger child. In contrast, imprinted genes that allow expression of maternally inherited alleles will tend to drive mechanisms to prevent the disproportionate utilization of resources by the fetus. A commonly cited example of this differential resource transfer in mice is the paternally expressed gene *Igf2* (insulin-like growth factor 2), which enhances fetal growth and placental nutrient transport capacity, and the maternally expressed *Igf2* receptor (*Igf2r*), which degrades excessive Igf2 protein.

Many of the effects of imprinted genes occur in the placenta, a crucial site for resource and nutrient transfer. For example, an overgrown placenta (hydatidiform mole) results when maternal imprints are missing. Additionally, in Silver-Russell syndrome (or Russell-Silver syndrome), a maternal uniparental disomy (both copies of a chromosome or partial chromosome are inherited from one parent), growth restriction is present. Similar effects are found in other cases of disordered imprinting. Preeclampsia, for example, in which disordered imprinting has been implicated, also demonstrates growth restriction in utero. Many of these diseases can be understood only within the context of imprinting as a common mechanism of parental conflict and manipulation of the phenotypic outcome of children.

Imprinting and cognition and behaviour after birth

Although only approximately 100 human genes are known to be subject to parent-of-origin effects, these prove to have tremendous implications for the development and eventual adult attributes of people, including cognitive and behavioral attributes. Imprinting effects, similar to other genetic expression patterns, may manifest as earlier age of onset effects or changes in severity.

In Turner syndrome, which affects females and is characterized by the deletion of one X chromosome (females normally carry two X chromosomes), a parent-of-origin effect may influence social and intellectual functioning. The affected female could have received her only X chromosome from her father (Xp) or her mother (Xm). Studies have suggested that neurodevelopment in Xp Turner syndrome patients is similar to that of healthy females, whereas neurodevelopment in Xm patients is altered. Other studies, however, have yielded conflicting results.

17

MATERNAL EFFECTS, CYTOPLASMIC INHERITANCE AND ORGANELLAR GENETICS

MATERNAL EFFECT

A maternal effect is a situation where the phenotype of an organism is determined not only by the environment it experiences and its genotype, but also by the environment and genotype of its mother. In genetics, maternal effects occur when an organism shows the phenotype expected from the genotype of the mother, irrespective of its own genotype, often due to the mother supplying messenger RNA or proteins to the egg. Maternal effects can also be caused by the maternal environment independent of genotype, sometimes controlling the size, sex, or behaviour of the offspring. These adaptive maternal effects lead to phenotypes of offspring that increase their fitness. Further, it introduces the concept of phenotypic plasticity, an important evolutionary concept. It has been proposed that maternal effects are important for the evolution of adaptive responses to environmental heterogeneity.

Maternal effects reflect the influence of the mother's gene on developing tissues. Many important characteristics of both animal and plants show maternal effects of which some examples axe described next.

COILING OF SNAIL SHELLS (LIMNAEA PEREGRA)

One of the earliest and classical examples of a maternal effect is that of the direction of coiling in shells of the water snail Limnaea peregra. In this snail, the shell is spirally coiled. Usually the direction of coiling of the shell is clockwise if viewed from the top of the shell. This type of coiling is called dextral. However, in some snails the coiling of shell is anticlockwise. This type of coiling is sinistral.

The direction of shell coiling of both types of snail is governed by genotype of the female parent and not by their own genotype. Further investigation suggests that coiling depends upon the early clearage in the zygote.

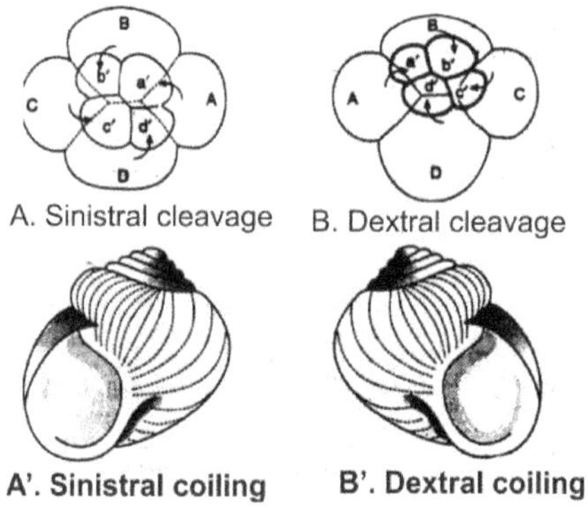

Fig. 1. A. Sinistral coiling of shell and B. Dextral coiling of shell

If the mitotic spindle is tilted to left (Fig. 1) of the median line of zygote, the successive cleavages will produce a spiral to left (sinistral) and if the orientation of spindle is tilted to the right of the median line of zygote, the successive cleavages will produce a spiral to right (dextral). The spindle orientation is controlled by the genotype of oocyte from which the egg develops.

In a cross between dextral 9 (female) snail and sinistral (male), all the Fi progeny have dextral coils like their mother and also indicates that dextral character (RR) is dominant over sinistral coiling (rr).

However, in the F_1 x F_1 cross (i.e., inbreeding or self fertilisation) all the F_2 snails are also dextral. The F_3 progeny from F_2 individuals with the genotype RR and R_r will show dextral coiling while those from rr F_2 individual will exhibit sinistral coiling of their shell; this produces the typical 1 : 3 ratio in F_3 generation.

In reciprocal cross between dextral B& (male) and sinistral @& (female), all the F_x progeny have sinistral coiling (Rr) instead of dextral coiling. In this case, F_1 x F_1 cross, all the F_2 snails are, again, dextral. This F_3 progenies from F_2 also exhibit the typical monohybrid ratio of 1 : 3.

MATERNAL INHERITANCE IN DROSOPHILA:

Several examples of maternal effects are known in drosophila:

i. Abnormal growth in the head region of Drosophila melanogaster was produced sporadically in a sample from a wild population collected at Acahuizotla, Mexico. Development of abnormal growth in the head region is called Tumorous head (Tu – h). Tu – h is governed by two major genes. But the frequency of tumour development in progeny is markedly influenced by the maternal effect.

   When a cross is made between a normal female fly and a male fly with head tumour, less than 1% of the progeny exhibit head tumour. In contrast, when a reciprocal cross is made between a female having head tumour and a normal male, about 30% of the progeny show tumour development.

ii. In Drosophila, fertility and survival are occasionally influenced by maternal effect. A recessive nuclear gene, grandchild-less, affects the fertility of progeny in Drosophila subobscura. In this fly a female homozygous for which is fertile but all her offspring's are sterile.

The reason of this effect arises from cytoplasmic dependent pathway for the development of many organs, the egg cytoplasm formed by a female fly is not uniform and various parts of the egg appear to be specifically assigned for the formation of different tissues. Thus the fate of Drosophila germ cells to produce either various or testes is determined early in development.

A daughterless (da) gene in Drosophila causes death of all XX zygotes derived from eggs of da da females. It is reported that the cytoplasm of eggs of da da females affects the two X-chromosomes and does not inhibit the development of the female phenotype.

It should be noted that the action of da is determined by the genotype of the female producing egg and not by the genotype of the egg cells themselves. It is also reported that interaction of da gene product is evident if the zygote is XX but there is no such action affecting the survival of XY zygote.

Some genes in Drosophila have lethal effects on male embryos. A recessive gene, sonless, leads to the absence of males in the progeny of females homozygous for this gene. Another gene, abnormal oocyte, reduces the frequency of males in the progeny.

These effects on the frequency of male progeny are produced due to effect of egg cytoplasm on the survival of XY zygote. Thus the genes for daughterless and sonless produce typical uniparental inheritance. These uniparental inheritance patterns depend on the nuclear genotype of the female parent.

MATERNAL INHERITANCE IN AMBYSTOMA:

Maternal inheritance has also been studied in Axololt of Ambystoma maxicanum. In this animal there is a lethal gene 'O'. It is a recessive gene. But its dominant gene is '+'. The heterozygous individuals with genotype '+O' are completely normal.

When two heterozygous individuals with genotype '+02 are crossed, a homozygous offspring with genotype '00' may be produced. The homozygous individual ('OO') develop normally in the early stage but in later life show a slight retardation of growth.

Their regeneration capacity is reduced to great extent and the homozygous males (OO) are sterile and their testes are poorly developed and spermatogenesis does not produce beyond spermatogonial stages. On the other hand, the homozygous females with genotype 'OO' produce eggs.

The eggs are capable of fertilisation and normal clearage but at the onset of gastrulation, further development is retarded and embryos normally die. This abnormal course of development is affected by the maternal effect, i.e., the growth is resulted by the genotype of the mother irrespective to that of the offspring.

If the maternal genes are recessive, i.e., 'OO', the growth of the offspring even with genotypes '+O' is retarded. According to Biggs and Justus (1967), in the egg of homozygous female 'OO' a protein-like substance called corrective factor is absent.

This protein-like substance is necessary for normal growth and development of the offspring and is synthesised by normal genes '++' of homozygous or heterozygous normal females ('++' or '+O') during oogenesis.

MATERNAL INHERITANCE OF EYE PIGMENTATION IN WATER FLEA AND FLOUR MOTH:

The maternal inheritance has also been observed in case of water flea (Gammarus sp.) and the flour moth (Ephestia kuhniella). The normal colour of both the invertebrates is dark due to the dominant gene (AA or KK) in which the dominant gene A or K directs the production of a hormone-like substance called Kynurenine which is involved in the pigment synthesis.

The recessive gene (i.e., a or k) fails to direct the synthesis of kynurenine. In absence of kynurenine, colour of eye becomes light.

Therefore the recessive mutants do not possess pigment in the eye and have the genotype aa or Kk. When a cross is made between a heterozygous male (Aa or Kk) and a double recessive female (aa or kk), only half of the larvae show dark pigment in the eye. Again, when a cross is made between a heterozygous female and a double recessive male, all larva are with dark eye. But on reaching the adult stage, half of the progenies having the genotype 'aa' or 'kk' become light-eyed.

This indicates that some kynurenine molecules diffuse from the 'Aa' mother into all young (larvae) enabling them to produce pigments regardless of their genotype. But the aa progeny is unable to synthesize kynurenine and, consequently, develops light eye as the kynurenine obtained from mother is used up. This example suggests an ephemeral type of maternal inheritance.

MATERNAL INHERITANCE IN MAMMALS

Maternal effects in mammals are very prominent because mammalian mothers may affect the development of their offspring's not only through ovum cytoplasm but also through the uterine environment. Effects of the maternal Rh blood group on the developing foetus in humans and that of maternal genotype for hair loss gene on the survival of young embryos in mice are the examples of maternal effect.

Other examples of maternal effects in human are: the embryonic defects caused by maternal diabetes and maternal phenylketonuria, a maternal influence on left-or right-handedness and a maternal effect on body weight.

MATERNAL INHERITANCE IN PLANTS

Cytoplasmic male sterility in cross-pollinating plants is a classic example of maternal inheritance. Plants which are unable to produce functional pollen but possess normal fertile female structure aired known as male sterile.

Thus male sterility is characterised by nonfunctional pollen grains. This occurs in many flowering plants. In maize, wheat, sugar-beets, onions and some other crop plants, fertility is controlled, at least in part—by cytoplasmic factors and results in male sterility.

This type of male sterility is referred to as cytoplasmic male sterility. Nuclear gene did not control this type of sterility, rather it is transmitted from generation to generation through egg cytoplasm. In other plants, however, male sterility is controlled entirely by nuclear genes. But over present discussion will be focused on only cytoplasmic male sterility in relation to maternal inheritance in plants.

Cytoplasmic male sterility is determined by cytoplasmic factors. Since the bulk of cytoplasm of zygote is contributed by the egg cell and the pollen tube containing male gametes contributes negligible or no cytoplasm, sterility factors present in the cytoplasm of egg cell will be transmitted to the offspring which would always be male sterile.

A case of cytoplasmic male sterility was discovered and carefully analysed by M. M. Rhoades (1933) in maize. He observed a male sterile plant in maize. In this plant pollens are aborted in the anther. The male sterile plant is produced when an egg cell containing cytoplasmic male sterility factor is fertilised by pollen from normal male fertile plants.

It indicates that male sterility is contributed by the cytoplasm of female parent. It is also- observed that when a male sterile female plant is crossed with wide range of fertile males, all progenies axe male sterile in the subsequent generations.

In maize, three distinct male sterile source (Cms) cytoplasm's axe known which are designated as Cms-T, Cms-C and Cms-S. The normal male fertile cytoplasm is known as N- cytoplasm. Each of the three Cms cytoplasm's shows strict maternal inheritance—even when all chromosomes axe replaced from male sterile plants by a male fertile source through repeated backcrossing.

Even then male sterility characteristics could not be avoided and the characteristics till persists. It indicates that if the character is present on chromosome as nuclear gene, then male sterility could be eliminated by repeated backcrossing.

Therefore, it is confirmed that male sterility is not controlled by nuclear gene, i.e., nuclear gene has no influence on cytoplasmic male sterility. In rare cases, male sterile plants produce a few fertile pollen grains.

When reciprocal crosses axe made between male sterile parent (with fertile pollen) and normal male fertile (female), the progeny is found to be male fertile. Such cases confirm maternal inheritance of male sterility.

Recent studies have shown that cytoplasmic male sterility is controlled by either some unique polypeptide produced by mitochondria or by some plasmid-like elements which axe not found in the mitochondria of normal cytoplasm.

Cytoplasmic male sterility due to S-cytoplasm is different from the cytoplasmic male sterility due to T-cytoplasm in several ways. The mt DNA of S-cytoplasm contains two unique plasmid-like DNA fragments called SF (Mol. wt. 3.45 x 10^6) and S-S (mol.wt.4.10 x 10^6).

These plasmid-like DNAs are not found in the isolated DNA of chloroplast or nuclei. Therefore, these plasmid-like DNAs are the characteristic of mitochondrial DNA of S-cytoplasm.

These plasmid-like DNAs are also absent in the mitochondrial DNA of N- cytoplasm (Normal-fertile), T-cytoplasm (male sterile) as well as C-cytoplasm (male sterile). Hence it has been suggested that such plasmid- like DNAs in mitochondria are responsible for causing male sterility.

T and C-cytoplasm of male sterile plant is comparatively stable and irreversible. It means that they never give rise to fertile cytoplasm even by applying mutagens.

CYTOPLASMIC INHERITANCE

Mitochondria and chloroplast are two organelles present in cells which contain DNA other than chromosomal DNA. These organellar DNA carries genetic information and works independently or in collaboration with nuclear DNA (chromosomal DNA). Inheritance of characteristics from generation to generation by extrachromosomal /cytoplasmic/ organelle DNA is called cytoplasmic inheritance. There are a large number of examples which show the involvement of cytoplasmic DNA in controlling heredity traits of the organisms. Hence, they are also known as cytoplasmic heredity units or cytoplasmic genes.

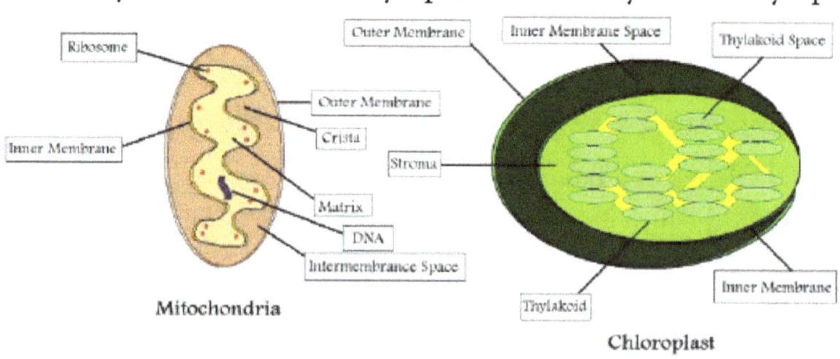

Fig. 2: Mitochondria and Chloroplast

These plasma genes are mostly received by the egg cytoplasm rather than the sperm cytoplasm. Hence, cytoplasmic inheritance is considered as a maternal inheritance phenomenon which influences the heredity characters. Though the cytoplasmic inheritance contributes to decide the characters of the offspring, reciprocal crosses do not result in the same phenotypes.

GENETIC MATERNAL EFFECT

Maternal effect is a situation which determines the phenotype of an offspring by the genotype of its mother, independent of the offspring genotype and environmental effect. In other words, the maternal effect is the casual influence of the maternal genotype on the phenotype of the offspring regardless of its genotype. It occurs due to the specific mRNA and proteins supplied by the mother to the zygote during embryo development. In many organisms, the embryo is initially inactive for the transcription. Hence, the supply of mRNA and proteins from the maternal side is important. The maternal effect does not arise due to heredity units. It arises totally due to these molecules received from the maternal supply. Due to these maternal effects, two offspring may sometimes differ phenotypically from each other though they possess the same genotype. One individual may resemble the mother parent.

The properties of cytoplasm are mainly governed by nuclear genes. Thus, the maternal effect depends on the nuclear genes.

Maternal effect is an important process in ecology and evolution. It contributes to population dynamics, phenotypic plasticity, niche construction, life-history evolution and natural selection.

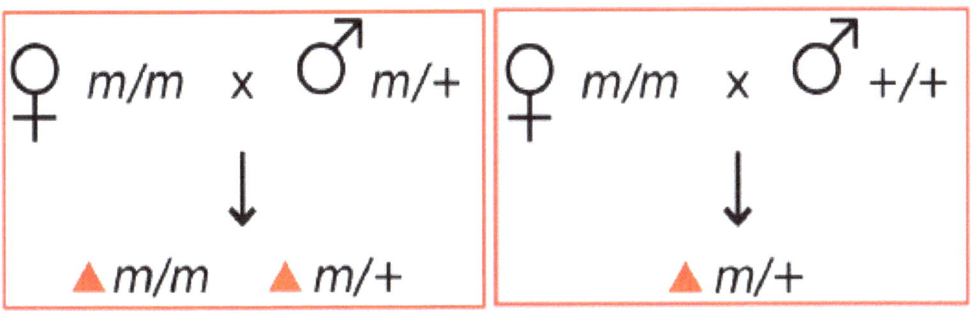

Fig. 3: Genetic crosses involving a maternal effect recessive mutation

Cytoplasmic Inheritance vs Genetic Maternal Effect

Chromosomal DNA is considered as the sole genetic material of a cell. However several cellular organelles (mitochondria, chloroplasts) possess DNA which can influence the traits of the offspring. Some maternal products in the cytoplasm are also involved in deciding the traits of an offspring. Cytoplasmic inheritance and genetic maternal effect are two such situations. These two phenomena are caused due to the genes or factors inherited from the mothers' egg to the zygote. Maternal effect is a result of mRNA and proteins (gene products) received from the cytoplasm of the mothers' egg. Cytoplasmic inheritance is a result of genetic material in the mitochondria or chloroplasts or infective viruses. This is the main difference between cytoplasmic inheritance and genetic maternal effect. Offspring inherits the maternal traits regardless of its own genotype and the genes due to both these phenomena.

ORGANELLAR GENETICS

In eukaryotes, DNA and genes also exist outside of the chromosomes found in the nucleus. Both the chloroplast and mitochondrion have circular chromosomes (Fig. 4). These organellar genomes are often present in multiple copies within each organelle. In most sexually reproducing species, organellar chromosomes are inherited from only one parent, usually the one that produces the largest gamete. Thus, in mammals, angiosperms, and many other organisms, mitochondria and chloroplasts are inherited only through the mother (maternally).

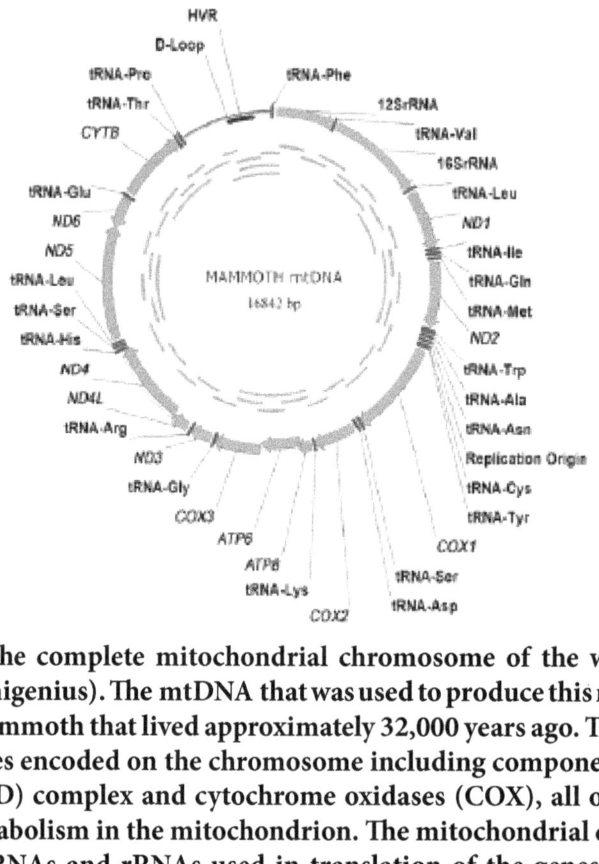

Fig. 4: A map of the complete mitochondrial chromosome of the woolly mammoth (Mammuthus primigenius). The mtDNA that was used to produce this map was obtained from tissue of a mammoth that lived approximately 32,000 years ago. The map shows the position of enzymes encoded on the chromosome including components of the NADH dehydrogenase (ND) complex and cytochrome oxidases (COX), all of which function during energy metabolism in the mitochondrion. The mitochondrial chromosome also encodes various tRNAs and rRNAs used in translation of the genes encoded on this chromosome. Other proteins required by the mitochondrion are encoded in the nuclear genome, and are translated in the cytoplasm and imported into the organelle. Circular organellar chromosomes such one as this are typical of almost all eukaryotes

These organelles are likely the remnants of prokaryotic endosymbionts that entered the cytoplasm of ancient progenitors of today's eukaryotes (endosymbiont theory). These endosymbionts had their own, circular chromosomes, like most bacteria that exist today. Chloroplasts and mitochondria typically have circular chromosomes that behave more like bacterial chromosomes than eukaryotic chromosomes, i.e. these organellar genomes do not undergo mitosis or meiosis.

Cell Organelles

More than 8.7 million species are living on the planet. Every single species is composed of a cell and it includes both single-celled and multicellular organisms.

The cells provide shape, structure and carries out different types of functions to keep the entire system active. The cell contains different functional structures which are collectively called Organelles, and they are involved in various cellular functions.

A cell is the basic structural and functional unit of a living organism. According to cell theory postulates, a cell is the basic building block of life, which makes anything alive and is self-sufficient to carry out all the fundamental functions of an organism.

CELL ORGANELLES

The cellular components are called cell organelles. These cell organelles include both membrane and non-membrane bound organelles, present within the cells and are distinct in their structures and functions. They coordinate and function efficiently for the normal functioning of the cell. A few of them function by providing shape and support, whereas some are involved in the locomotion and reproduction of a cell. There are various organelles present within the cell and are classified into three categories based on the presence or absence of membrane.

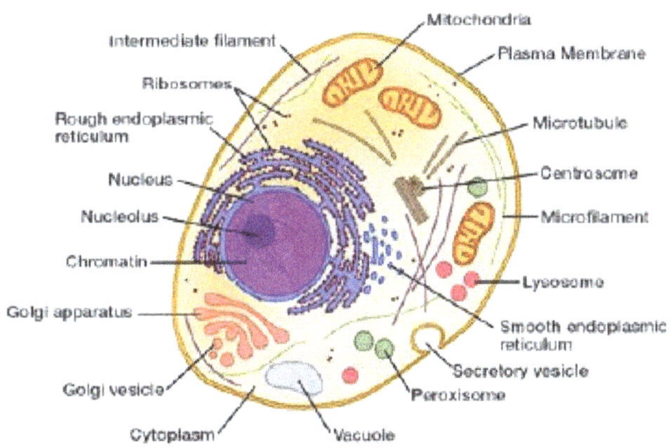

Fig. 5. Cell Organelles

Organelles without membrane: The Cell wall, Ribosomes, and Cytoskeleton are non-membrane-bound cell organelles. They are present both in prokaryotic cell and the eukaryotic cell.

Single membrane-bound organelles: Vacuole, Lysosome, Golgi Apparatus, Endoplasmic Reticulum are single membrane-bound organelles present only in a eukaryotic cell.

Double membrane-bound organelles: Nucleus, mitochondria and chloroplast are double membrane-bound organelles present only in a eukaryotic cell.

LIST OF CELL ORGANELLES AND THEIR FUNCTIONS

Plasma Membrane

The plasma membrane is also termed as a Cell Membrane or Cytoplasmic Membrane. It is a selectively permeable membrane of the cell, which is composed of a lipid bilayer and proteins.

The plasma membrane is present both in plant and animal cell, which functions as the selectively permeable membrane, by permitting the entry of selective materials in and out of the cell according to the requirement. In an animal cell, the cell membrane functions by providing shape and protects the inner contents of the cell. Based on the structure of the plasma membrane, it is regarded as the fluid mosaic model. According to the fluid mosaic model, the plasma membranes are subcellular structures, made of a lipid bilayer in which the protein molecules are embedded.

Cytoplasm

The cytoplasm is present both in plant and animal cells. They are jelly-like substances, found between the cell membrane and nucleus. They are mainly composed of water, organic and inorganic compounds. The cytoplasm is one of the essential components of the cell, where all the cell organelles are embedded. These cell organelles contain enzymes, mainly responsible for controlling all metabolic activity taking place within the cell and are the site for most of the chemical reactions within a cell.

Nucleus

The nucleus is a double-membraned organelle found in all eukaryotic cells. It is the largest organelle, which functions as the control centre of the cellular activities and is the storehouse of the cell's DNA. By structure, the nucleus is dark, round, surrounded by a nuclear membrane. It is a porous membrane (like cell membrane) and forms a wall between cytoplasm and nucleus. Within the nucleus, there are tiny spherical bodies called nucleolus. It also carries another essential structure called chromosomes.

Chromosomes are thin and thread-like structures which carry another important structure called a gene. Genes are a hereditary unit in organisms i.e., it helps in the inheritance of traits from one generation (parents) to another (offspring). Hence, the nucleus controls the characters and functions of cells in our body. The primary function of the nucleus is to monitor cellular activities including metabolism and growth by making use of DNA's genetic information. Nucleoli in the nucleus are responsible for the synthesis of protein and RNA.

Endoplasmic Reticulum

The Endoplasmic Reticulum is a network of membranous canals filled with fluid. They are the transport system of the cell, involved in transporting materials throughout the cell.

There are two different types of Endoplasmic Reticulum:

1. **Rough Endoplasmic Reticulum**– They are composed of cisternae, tubules, and vesicles, which are found throughout the cell and are involved with protein manufacture.
2. **Smooth Endoplasmic Reticulum**– They are the storage organelle, associated with the production of lipids, steroids, and also responsible for detoxifying the cell.

Mitochondria

Mitochondria are called the powerhouses of the cell as they produce energy-rich molecules for the cell. The mitochondrial genome is inherited maternally in several organisms. It is a double membrane-bound, sausage-shaped organelle, found in almost all eukaryotic cells.

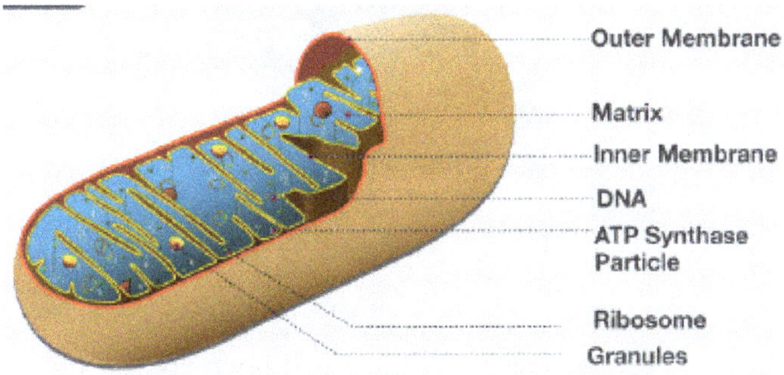

Fig. 7. Mitochondria

The double membranes divide its lumen into two distinct aqueous compartments. The inner compartment is called 'matrix' which is folded into cristae whereas the outer membrane forms a continuous boundary with the cytoplasm. They usually vary in their size and are found either round or oval in shape. Mitochondria are the sites of aerobic respiration in the cell, produces energy in the form of ATP and helps in the transformation of the molecules.

For instance, glucose is converted into adenosine triphosphate – ATP. Mitochondria have their own circular DNA, RNA molecules, ribosomes (the 70s), and a few other molecules that help in protein synthesis.

Plastids

Plastids are large, membrane-bound organelles which contain pigments. Based on the type of pigments, plastids are of three types:

- **Chloroplasts**– Chloroplasts aredouble membrane-bound organelles, which usually vary in their shape – from a disc shape to spherical, discoid, oval and ribbon. They are present in mesophyll cells of leaves, which store chloroplasts and other carotenoid pigments. These pigments are responsible for trapping light energy for photosynthesis. The inner membrane encloses a space called the stroma. Flattened disc-like chlorophyll-containing structures known as thylakoids are arranged in a stacked manner like a pile of coins. Each pile is called as granum (plural: grana) and the thylakoids of different grana are connected by flat membranous tubules known as stromal lamella. Just like the mitochondrial matrix, the stroma of chloroplast also contains a double-stranded circular DNA, 70S ribosomes, and enzymes which required for the synthesis of carbohydrates and proteins.

- **Chromoplasts**– The chromoplasts include fat-soluble, carotenoid pigments like xanthophylls, carotene, etc. which provide the plants with their characteristic color – yellow, orange, red, etc.

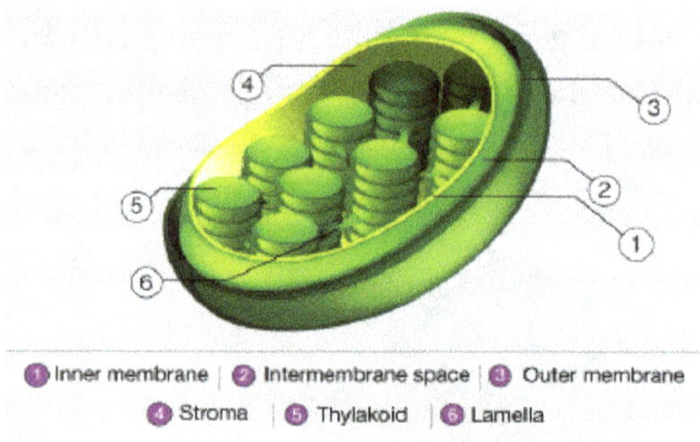

Fig. 8. Chloroplasts

- **Leucoplasts**– Leucoplasts are colorless plastids which store nutrients. Amyloplasts store carbohydrates (like starch in potatoes), aleuroplasts store proteins, and elaioplasts store oils and fats.

Ribosomes are nonmembrane-bound and important cytoplasmic organelles found in close association with the endoplasmic reticulum. Ribosomes are found in the form of tiny particles in a large number of cells and are mainly composed of 2/3rd of RNA and 1/3rd of protein. They are named as the 70s (found in prokaryotes) or 80s (found in eukaryotes) The letter S refers to the density and the size, known as Svedberg's Unit. Both 70S and 80S ribosomes are composed of two sub-units. Ribosomes are either encompassed within the endoplasmic reticulum or are freely traced in the cell's cytoplasm. Ribosomal RNA and Ribosomal proteins are the two components that together constitute ribosomes. The primary function of the ribosomes includes protein synthesis in all living cells that ensure the survival of the cell.

Fig. 9. Ribosomes

Golgi Apparatus

Golgi Apparatus also termed as Golgi Complex. It is a membrane-bound organelle, which is mainly composed of a series of flattened, stacked pouches called cisternae. This cell organelle is primarily responsible for transporting, modifying, and packaging proteins and lipid to targeted destinations. Golgi Apparatus is found within the cytoplasm of a cell and are present in both plant and animal cells.

Microbodies

Microbodies are membrane-bound, minute, vesicular organelles, found in both plant and animal cell. They contain various enzymes and proteins and can be visualized only under the electron microscope.

Cytoskeleton

It is a continuous network of filamentous proteinaceous structures that run throughout the cytoplasm, from the nucleus to the plasma membrane. It is found in all living cells, notably in the eukaryotes. The cytoskeleton matrix is composed of different types of proteins that can divide rapidly or disassemble depending on the requirement of the cells. The primary functions include providing the shape and mechanical resistance to the cell against deformation, the contractile nature of the filaments helps in motility and during cytokinesis.

Cilia and Flagella

Cilia are hair-like projections, small structures, present outside the cell wall and work like oars to either move the cell or the extracellular fluid. Flagella are slightly bigger and are responsible for the cell movements. The eukaryotic flagellum structurally differs from its prokaryotic counterpart. The core of the cilium and flagellum is called a axoneme, which contains nine pairs of gradually arranged peripheral microtubules and a set of central microtubules running parallel to the axis. The central tubules are interconnected by a bridge and are embedded by a central sheath. One of the peripheral microtubular pairs is also interconnected to the central sheath by a radial spoke. Hence there is a total of 9 radial spokes. The cilia and flagella emerge from centriole-like structures called basal bodies.

Centrosome and Centrioles

The centrosome organelle is made up of two mutually perpendicular structures known as centrioles. Each centriole is composed of 9 equally spaced peripheral fibrils of tubulin protein, and the fibril is a set of interlinked triplets. The core part of the centriole is known as a hub and is proteinaceous. The hub connects the peripheral fibrils via radial spoke, which is made up of proteins. The centrioles from the basal bodies of the cilia and flagella give rise to spindle fibres during cell division.

Vacuoles

Vacuoles are mostly defined as storage bubbles of irregular shapes which are found in cells. They are fluid-filled organelles enclosed by a membrane. The vacuole stores the food or a variety of nutrients that a cell might need to survive. In addition to this, it also stores waste products. The waste products are eventually thrown out by vacuoles. Thus, the rest of the cell is protected from contamination. The animal and plant cell have different size and number of vacuoles. Compared to the animals, plant cell have larger vacuoles.

18

MUTATIONS: MORPHOLOGICAL LEVEL (INCLUDING LETHAL MUTATIONS)

Mutations are changes in the genetic sequence, and they are a main cause of diversity among organisms. These changes occur at many different levels, and they can have widely differing consequences. In biological systems that are capable of reproduction, we must first focus on whether they are heritable; specifically, some mutations affect only the individual that carries them, while others affect all of the carrier organism's offspring, and further descendants. For mutations to affect an organism's descendants, they must:

1. occur in cells that produce the next generation, and
2. affect the hereditary material. Ultimately, the interplay between inherited mutations and environmental pressures generates diversity among species.

Although various types of molecular changes exist, the word "mutation" typically refers to a change that affects the nucleic acids. In cellular organisms, these nucleic acids are the building blocks of DNA, and in viruses they are the building blocks of either DNA or RNA. One way to think of DNA and RNA is that they are substances that carry the long-term memory of the information required for an organism's reproduction. This article focuses on mutations in DNA, although we should keep in mind that RNA is subject to essentially the same mutation forces.

If mutations occur in non-germline cells, then these changes can be categorized as somatic mutations. The word somatic comes from the Greek word *soma* which means "body", and somatic mutations only affect the present organism's body. From an evolutionary perspective, somatic mutations are uninteresting, unless they occur systematically and change some fundamental property of an individual - such as the capacity for survival. For example, cancer is a potent somatic mutation that will affect a single organism's survival.

As a different focus, evolutionary theory is mostly interested in DNA changes in the cells that produce the next generation.

A mutation is a change in the DNA sequence of an organism. Mutations can result from errors in DNA replication during cell division, exposure to mutagens or a viral infection. Germline mutations (that occur in eggs and sperm) can be passed on to offspring, while somatic mutations (that occur in body cells) are not passed on.

Mutations in the structure of genes can be classified as Small-scale Mutationsand Large Scale Mutations.

Small-scale mutations are types of gene mutations, such as those affecting a small gene in one or a few nucleotides, including:

POINT MUTATIONS

A mutation is said to be punctual when it touches one or more nucleotides of the same gene.

Substitution mutations

- **Missense mutations:** This point mutation results in the replacement of one nucleotide by another. In some cases, this change causes a change in the amino acid encoded, which may or may not have an impact on the function of the protein produced by the gene in the case of a gene encoding, or the affinity for a transcription factor, in the case of a promoter region of the DNA.

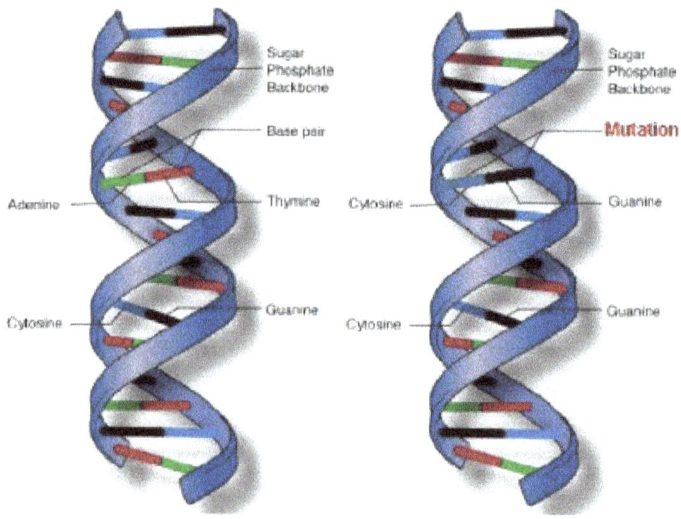

Fig. 1 Point mutations

- **Nonsense mutation:** In this instead of substituting one amino acid for another, however, the altered DNA sequence prematurely signals the cell to stop building a protein. This type of mutation results in a shortened protein that may function improperly or not at all.
- **Silent mutations:** These are mutations that do not alter the sequence of a protein because of the redundancy of the genetic code (the new triplet codes for the same

amino acid as the original triplet), or because it affects an area not coding DNA or an intron.

The insertions and deletions are the type of mutations and are the two types of mutations called frame-shift. Addition or deletion of nucleotides is not a multiple of 3 will cause a change of reading frame of the genetic code.

- Insertions add one or more extra nucleotides into the DNA. An insertionchanges the number of DNA bases in a gene by adding a piece of DNA.
- Deletions mean removing one or more nucleotides from the DNA. Like insertions, these mutations can alter the reading frame of the gene.

LARGE SCALE MUTATIONS

Chromosomal Mutations

It can be a loss or gain of chromosomes:These mutations change from one generation to the next, they correspond to the high repetition of some triplets at the DNA level

Somatic or Germline mutations

We speak of germ-line mutation or de-novo mutation when the mutation involves the DNA of stem cells from a gamete. In this case, the embryo will carry the mutation without any of its parents are possessed in his genetic heritage. This type of mutation occurs during the formation of life or the gametes of one parent (sperm or ovum).

MORPHOLOGICAL LEVEL (INCLUDING LETHAL MUTATIONS)

Morphological mutants affect the outward appearance of an individual. Plant height mutations could changes a tall plant to a short one, or from having smooth to round seeds. Biochemical mutations have a lesion in one specific step of an enzymatic pathway. For bacteria, biochemical mutants need to be grown on a media supplemented with a specific nutrient. Such mutants are called auxotrophs. Often though, morphological mutants are the direct result of a mutation in a biochemical pathway. In humans, albinism is the result of a mutation in the pathway from converts the amino acid tyrosine to the skin pigment melanin. Similarly, cretinism results when the tyrosine to thyroxine pathway is mutated. Therefore, in a strict genetic sense, if appropriate experiments are performed, a morphological mutation can be explained at the biochemical level.

For some mutations to be expressed, the individual needs to be placed in a specific environment. This is called the restrictive condition. But if the individual grow in any other environment (permissive condition), the wild type phenotype is expressed. These are called conditional mutations. Mutations that only expressed at a specific temperature (temperature sensitive mutants), usually elevated, can be considered to be conditional mutations.

The chicken's comb, which in the wild type is a single serrated blade on top of the head, has taken on a range of new shapes in domestic chickens (Fig. 2). These alternate comb forms have long been known to have a simple heritable basis, serving as major traits for the earliest demonstrations of Mendelian inheritance in animals and yielding a first example of epistatic interaction between genes. Andersson and colleagues set out to identify the genetic basis for the three major variants of the comb, Pea-comb (a smaller comb typically composed of three knobbed ridges), Rose-comb (flattened with papillae towards the beak and tapered to the back), and duplex-combs (either a full comb duplication or small paired horns), providing a systematic set in which to discern whether any particular type of mutation drives alteration of form. The commonalities uncovered are striking.

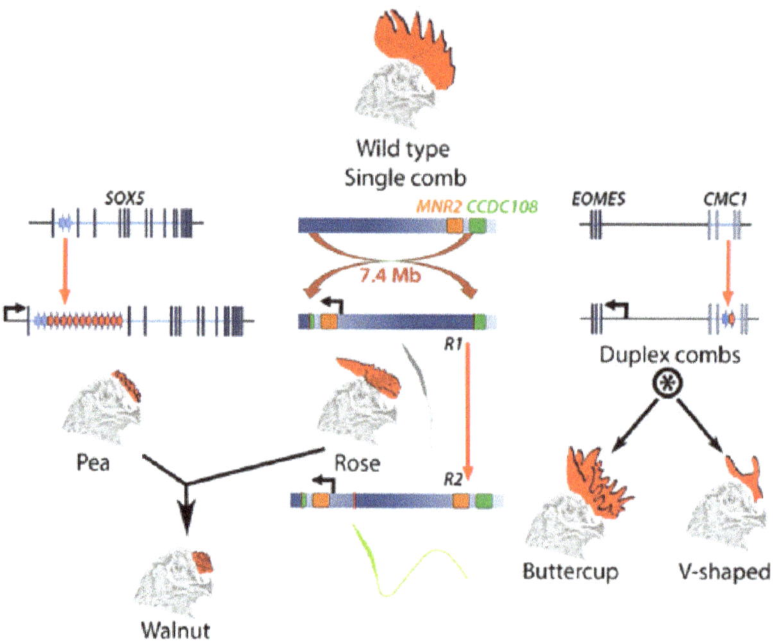

Fig 2. Major comb variants and their genetic basis.

All three comb variants are underlain by regulatory mutations that are structural, rather than single nucleotide changes, and each causes ectopic expression of a transcription factor. Pea-comb is caused by an approximately 30-fold expansion of a pre-existing tandem duplication in noncoding sequence at *SOX5*; Rose-comb by a large inversion, which induces expression of *MNR2*; and the duplex-comb phenotypes are now revealed to be a result of a tandem duplication in an intron of *CMC1*, which triggers expression of the neighbouring gene *EOMES*.

These acquisitions of new expression domains at the prospective comb region occur despite there being very little new sequence generated by the mutations, the bulk of which constitute amplification or rearrangement of existing sequences. It will be interesting to determine the gene regulatory mechanisms underlying the effect of the tandem expansions; whether these disrupt endogenous repressive elements, cause de novo formation of site-specific enhancers, or exert a more general locus-wide effect to permit the action of previously cryptic enhancers. The resulting ectopic expression of each transcription factor presumably amplifies the effect of the causative mutation into altered expression of many genes, thereby modifying the intercellular signalling that controls comb outgrowth.

Variation in domestic animals has been used as a guide to understand variation between species since the beginning of evolutionary thinking. However, sheltered from the full force of natural selection by human management, domesticated populations may be able to harbour crude mutations that would not be maintained in wild populations. Here, too, these comb studies have lessons, showing that further evolution of the original mutant alleles occurs either to reduce pleiotropic effects on fitness or to achieve finer tuning of the selected morphological phenotype. The former phenomenon is exemplified by Rose-comb, the original mutant allele of which causes sub-fertility, which has acquired a second rearrangement to repair this defect.

Further refinement of form is illuminated by the duplex-comb's two distinct shapes, which carry the same driving mutation, indicating that a second mutation arising at this site, or possibly a closely linked modifier allele, determines the difference between these morphologies. Taken together, this catalogue of mutations hints at the types of genomic change that tend to serve as the source of morphological variation within, and perhaps between, animal species.

Lethal mutations are also possible. As the term implies, the mutations lead to the death of the individual. Death does not have to occur immediately, it may take several months or even years. But if the expected longevity of an individual is significantly reduced, the mutation is considered a lethal mutation.

Wild type alleles typically encode a product necessary for a specific biological function. If a mutation occurs in that allele, the function for which it encodes is also lost. The general term for these mutations is loss-of-function mutations. The degree to which the function is lost can vary. If the function is entirely lost, the mutation is called a null mutation. If is also possible that some function may remain, but not at the level of the wild type allele. These are called leaky mutations.

Loss of function mutations are typically recessive. When a heterozygote consists of the wild-type allele and the loss-of-function allele, the level of expression of the wild type allele is often sufficient to produce the wild type phenotype. Genetically this would define the loss-of-function mutation as recessive. Alternatively, the wild type allele may not compensate for the loss-of-function allele. In those cases, the phenotype of the heterozygote will be equal to that of the loss-of-function mutant, and the mutant allele will act as a dominant.

Although it would be expected that most mutations would lead to a loss of function, it is possible that a new and important function could result from the mutation. In these cases, the mutation creates a new allele that is associated with a new function. Any heterozygote containing the new allele along with the original wild type allele will express the new allele. Genetically this will define the mutation as a dominant. This class of mutations are called gain-of-function mutations.

19

MUTATIONS: BIOCHEMICAL AND MICROBIAL GENETICS

A mutation is a heritable change in the DNA sequence of an organism. The resulting organism, called a mutant, may have a recognizable change in phenotype compared to the wild type, which is the phenotype most commonly observed in nature. A change in the DNA sequence is conferred to mRNA through transcription, and may lead to an altered amino acid sequence in a protein on translation. Because proteins carry out the vast majority of cellular functions, a change in amino acid sequence in a protein may lead to an altered phenotype for the cell and organism.

EFFECTS OF MUTATIONS ON DNA SEQUENCE

There are several types of mutations that are classified according to how the DNA molecule is altered. One type, called a point mutation, affects a single base and most commonly occurs when one base is substituted or replaced by another. Mutations also result from the addition of one or more bases, known as an insertion, or the removal of one or more bases, known as a deletion.

EFFECTS OF MUTATIONS ON PROTEIN STRUCTURE AND FUNCTION

Point mutations may have a wide range of effects on protein function (Fig. 1). As a consequence of the degeneracy of the genetic code, a point mutation will commonly result in the same amino acid being incorporated into the resulting polypeptide despite the sequence change. This change would have no effect on the protein's structure, and is thus called a silent mutation. A missense mutation results in a different amino acid being incorporated into the resulting polypeptide. The effect of a missense mutation depends on how chemically different the new amino acid is from the wild-type amino acid. The location of the changed amino acid within the protein also is important. For example, if the changed amino acid is part of the enzyme's active site, then the effect of the missense mutation may be significant. Many missense mutations result in proteins that are still functional, at least to some degree. Sometimes the effects of missense mutations may be only apparent under certain environmental conditions; such missense mutations are called conditional mutations. Rarely, a missense mutation may be beneficial. Under the right environmental conditions, this type of mutation may give the organism that harbors it a selective advantage. Yet another type of point mutation, called a nonsense mutation, converts a codon encoding an amino acid (a sense codon) into a stop codon (a nonsense codon). Nonsense mutations result in the synthesis of proteins that are shorter than the wild type and typically not functional.

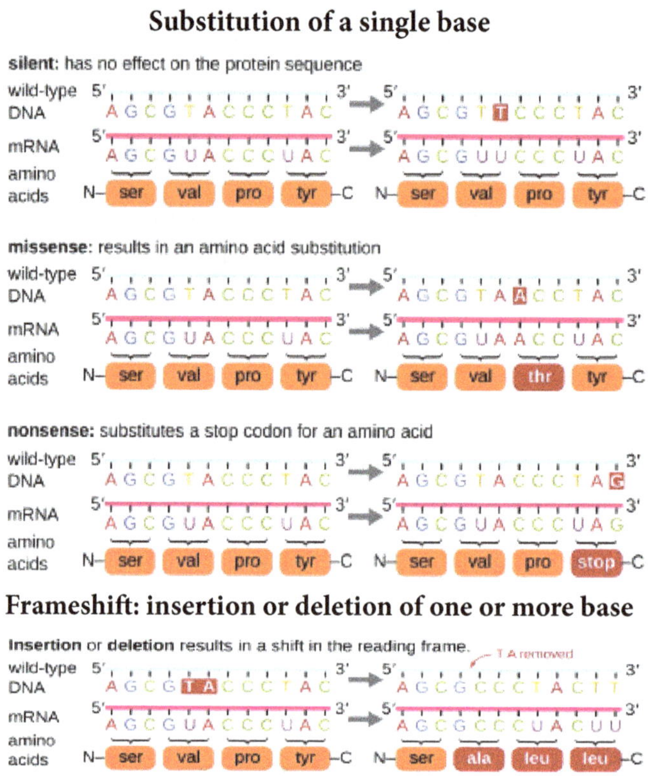

Fig. 1. Click for a larger image. Mutations can lead to changes in the protein sequence encoded by the DNA.

Deletions and insertions also cause various effects. Because codons are triplets of nucleotides, insertions or deletions in groups of three nucleotides may lead to the insertion or deletion of one or more amino acids and may not cause significant effects on the resulting protein's functionality. However, frameshift mutations, caused by insertions or deletions of a number of nucleotides that are not a multiple of three are extremely problematic because a shift in the reading frame results (Fig. 1). Because ribosomes read the mRNA in triplet codons, frameshift mutations can change every amino acid after the point of the mutation. The new reading frame may also include a stop codon before the end of the coding sequence. Consequently, proteins made from genes containing frameshift mutations are nearly always nonfunctional.

Various types of chemical mutagens interact directly with DNA either by acting as nucleoside analogs or by modifying nucleotide bases. Chemicals called nucleoside analogs are structurally similar to normal nucleotide bases and can be incorporated into DNA during replication (Fig. 2). These base analogs induce mutations because they often have different base-pairing rules than the bases they replace. Other chemical mutagens can modify normal DNA bases, resulting in different base-pairing rules. For example, nitrous acid deaminates cytosine, converting it to uracil. Uracil then pairs with adenine in a subsequent round of replication, resulting in the conversion of a GC base pair to an AT base pair. Nitrous acid also deaminates adenine to hypoxanthine, which base pairs with cytosine instead of thymine, resulting in the conversion of a TA base pair to a CG base pair.

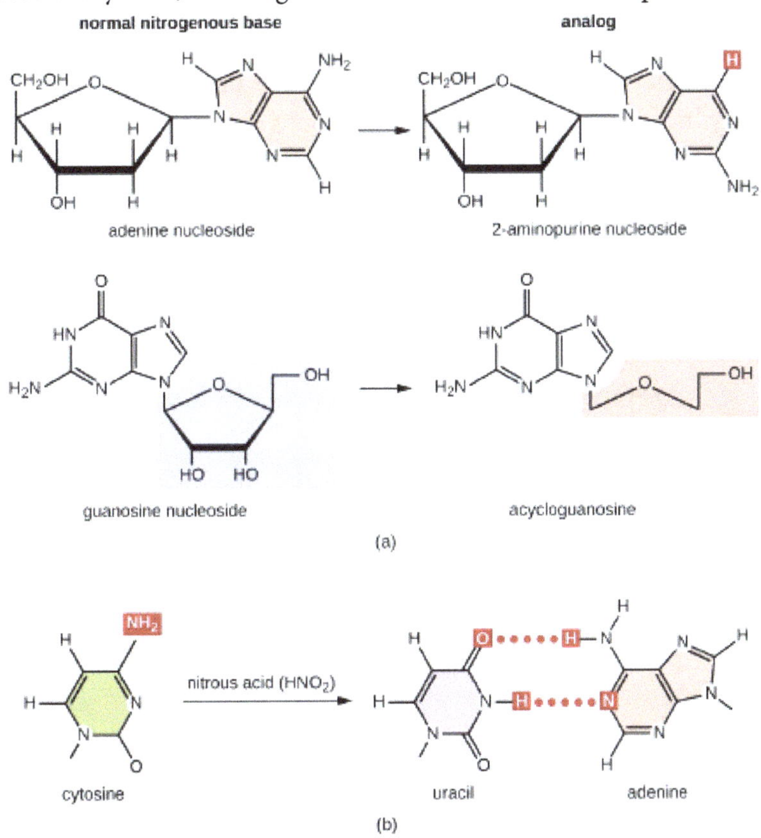

Fig. 2. Click for a larger image. (a) 2-aminopurine nucleoside (2AP) structurally is a nucleoside analog to adenine nucleoside, whereas 5-bromouracil (5BU) is a nucleoside analog to thymine nucleoside. 2AP base pairs with C, converting an AT base pair to a GC base pair after several rounds of replication. 5BU pairs with G, converting an AT base pair to a GC base pair after several rounds of replication. (b) Nitrous acid is a different type of chemical mutagen that modifies already existing nucleoside bases like C to produce U, which base pairs with A.

Chemical mutagens known as intercalating agents work differently. These molecules slide between the stacked nitrogenous bases of the DNA double helix, distorting the molecule and creating atypical spacing between nucleotide base pairs (Fig. 3). As a result, during DNA replication, DNA polymerase may either skip replicating several nucleotides (creating a deletion) or insert extra nucleotides (creating an insertion). Either outcome may lead to a frameshift mutation. Combustion products like polycyclic aromatic hydrocarbons are particularly dangerous intercalating agents that can lead to mutation-caused cancers. The intercalating agents ethidium bromide and acridine orange are commonly used in the laboratory to stain DNA for visualization and are potential mutagens.

RADIATION

Exposure to either ionizing or nonionizing radiation can each induce mutations in DNA, although by different mechanisms. Strong ionizing radiation like X-rays and gamma rays can cause single- and double-stranded breaks in the DNA backbone through the formation of hydroxyl radicals on radiation exposure (Fig. 4). Ionizing radiation can also modify bases; for example, the deamination of cytosine to uracil, analogous to the action of nitrous acid. Ionizing radiation exposure is used to kill microbes to sterilize medical devices and foods, because of its dramatic nonspecific effect in damaging DNA, proteins, and other cellular components.

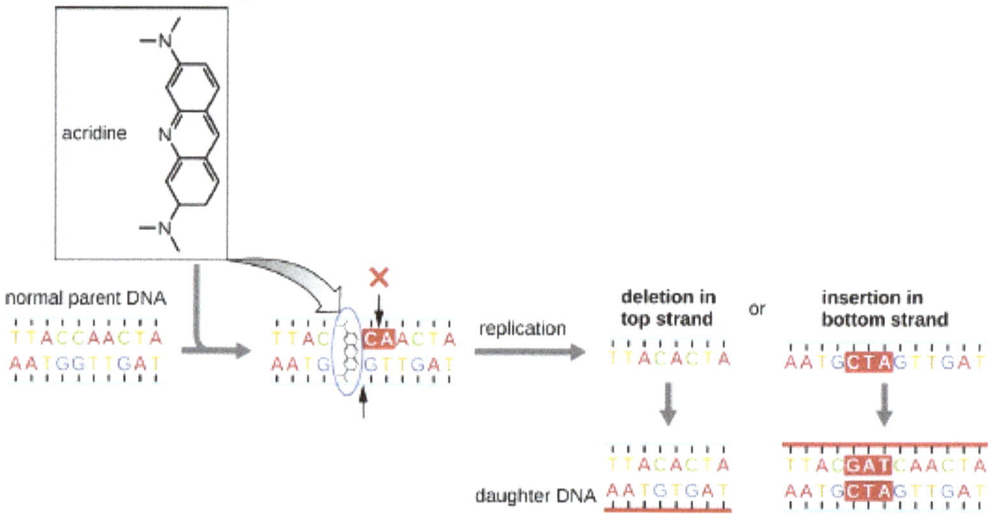

Fig. 3. Intercalating agents, such as acridine, introduce atypical spacing between base pairs, resulting in DNA polymerase introducing either a deletion or an insertion, leading to a potential frameshift mutation.

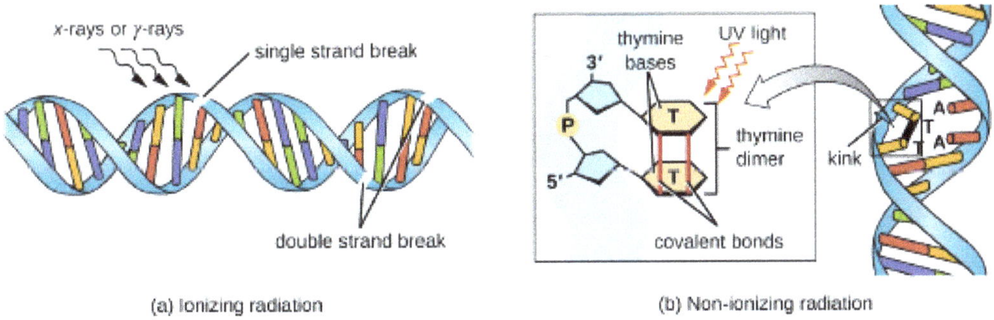

Fig. 4. (a) Ionizing radiation may lead to the formation of single-stranded and double-stranded breaks in the sugar-phosphate backbone of DNA, as well as to the modification of bases (not shown). (b) Nonionizing radiation like ultraviolet light can lead to the formation of thymine dimers, which can stall replication and transcription and introduce frameshift or point mutations.

Nonionizing radiation, like ultraviolet light, is not energetic enough to initiate these types of chemical changes. However, nonionizing radiation can induce dimer formation between two adjacent pyrimidine bases, commonly two thymines, within a nucleotide strand. During thymine dimer formation, the two adjacent thymines become covalently linked and, if left unrepaired, both DNA replication and transcription are stalled at this point. DNA polymerase may proceed and replicate the dimer incorrectly, potentially leading to frameshift or point mutations.

MOLECULAR BIOLOGY

20

STRUCTURE AND SYNTHESIS OF NUCLEIC ACIDS

Nucleic acids are large biomolecules that play essential roles in all cells and viruses. A major function of nucleic acids involves the storage and expression of genomic information. Deoxyribonucleic acid, or DNA, encodes the information cells need to make proteins. A related type of nucleic acid, called ribonucleic acid (RNA), comes in different molecular forms that play multiple cellular roles, including protein synthesis.

Nucleic acIDS naturally occurring chemical compound that is capable of being broken down to yield phosphoric acid, sugars, and a mixture of organic bases (purines and pyrimidines). Nucleic acids are the main information-carrying molecules of the cell, and, by directing the process of protein synthesis, they determine the inherited characteristics of every living thing. The two main classes of nucleic acids are deoxyribonucleic acid (DNA) and ribonucleic acid (RNA). DNA is the master blueprint for life and constitutes the genetic material in all free-living organisms and most viruses. RNA is the genetic material of certain viruses, but it is also found in all living cells, where it plays an important role in certain processes such as the making of proteins.

Although the nucleic acids were first discovered in 1868, by Friedrich Miescher working with pus cells obtained from discarded surgical bandages, it was not really until the early 1940s that the chemistry and biology of the nucleic acids were set on firm foundations. Basically, nucleic acids can be subdivided into two types: deoxyribonucleic acid (DNA) and ribonucleic acid (RNA). Both DNA and RNA have been shown to consist of three groups of molecules: pentose (5-carbon-atom) sugars; organic bases; and inorganic phosphate.

SUGARS

There are only two types of sugar present in nucleic acids, ribose which is present solely in RNA (hence its name) and deoxyribose which is present solely in DNA (again, the sugar gives rise to the name deoxyribonucleic acid). The chemical structures for these compounds are shown here:

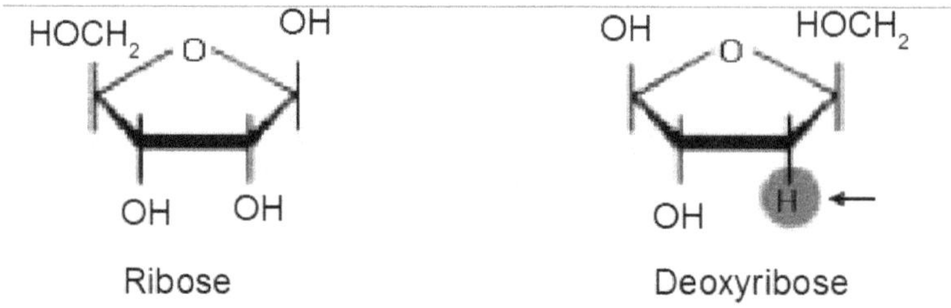

Fig. 1

The prefix 'deoxy' means 'without oxygen', and we can see from the structures that the only difference between them is the absence of an oxygen in the deoxyribose sugar (see shaded area). Both sugars contain 5 carbon atoms (pentose sugars) and for convenience we number these as shown in the next figure. The 'dash' or 'prime' (') on, for example, the 5 indicates the carbon in the ribose ring. The purine or pyrimidine rings are numbered without primes in order to distinguish them.

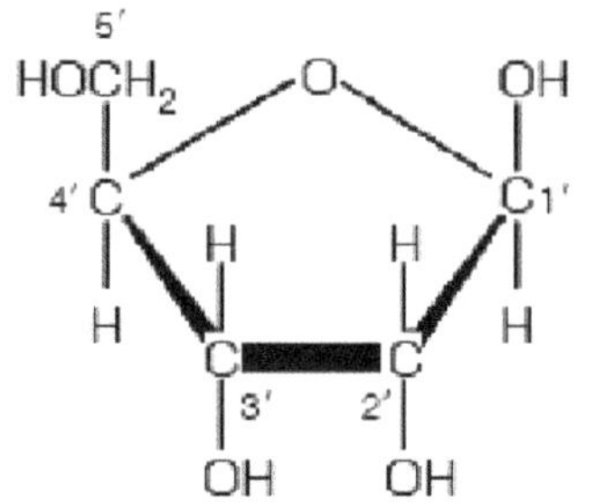

Fig. 2

Bases The nucleotide bases found in nucleic acids are related either to the purine ring system or to the pyrimidine ring system

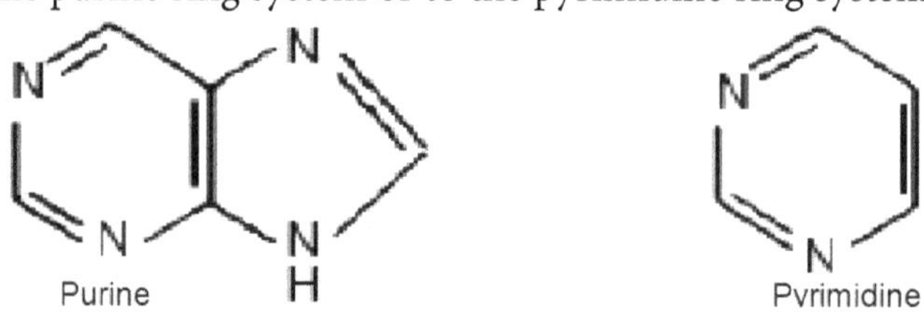

Fig. 3

In DNA we find principally four different bases: adenine (A), guanine (G), cytosine (C) and thymine (T). The first two are derived from purine whereas the remaining two are derived from pyrimidine. In RNA we find principally four different bases; adenine, guanine and cytosine as in DNA. The fourth base in RNA, however, is not thymine but instead the pyrimidine-derived base, uracil (U). The chemical structures for each of the five bases are shown below (Fig. 4)

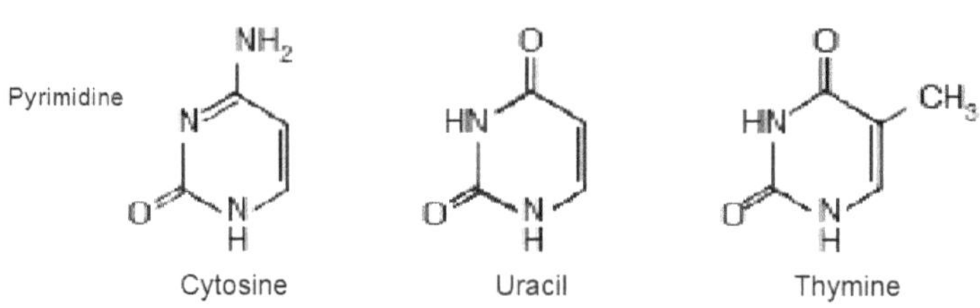

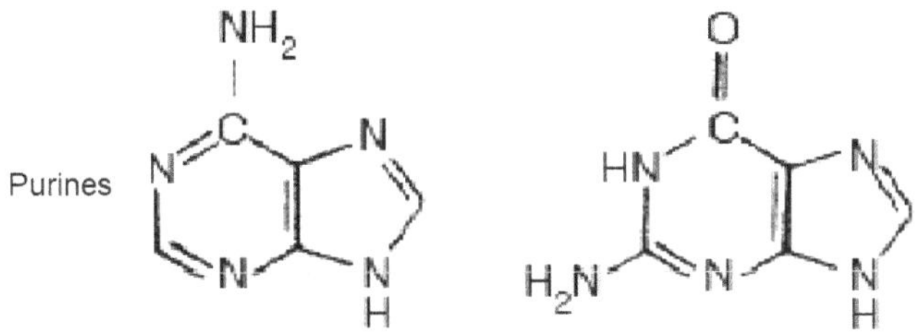

Adenine Guanine

Fig. 4

In addition to these major bases there is also a large range of so called minor bases which occur less frequently than the others, e.g. 5-methyl cytosine.

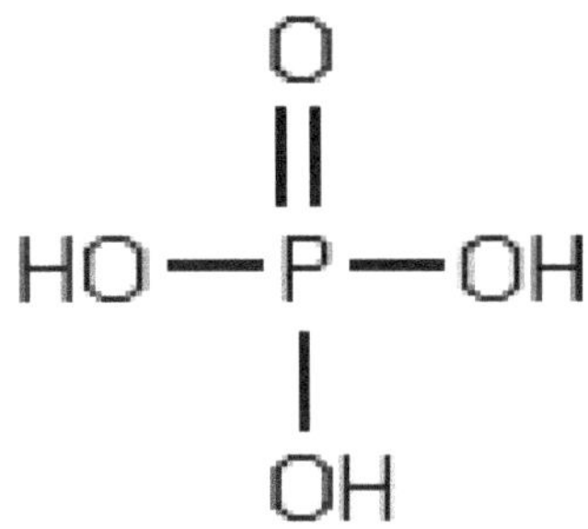

Fig 5

INORGANIC PHOSPHATE

There are phosphate residues in nucleic acids and they are of the type derived from phosphoric acid, the structure of which is shown in Fig 5.

BUILDING NUCLEIC ACIDS FROM THEIR BUILDING BLOCKS

When any one of the bases is joined to either one of the two sugar molecules, we have a compound known as a nucleoside. If the sugar residue is ribose then we have a ribonucleoside, whereas if it is deoxyribose then we have a deoxyribonucleoside.

The bond linking these structures is known as a glycoside bond. Addition of a phosphate group to the sugar residue of a nucleoside molecule produces a different molecule called a nucleotide. Examples of various nucleotides are shown in Fig. 6.

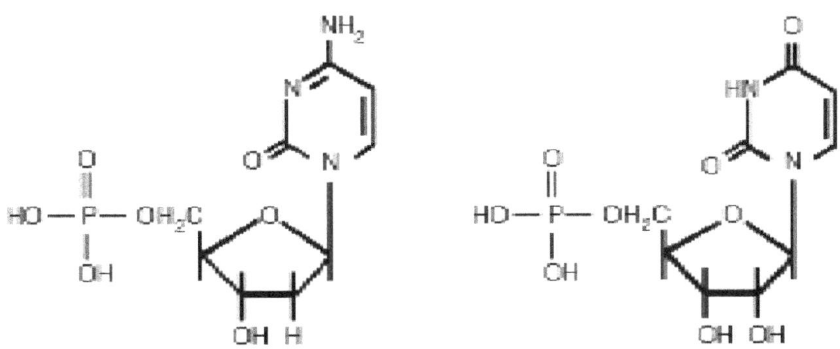

Deoxycytidine 5-monophosphate Uridine 5-monophosphate

Fig. 6

Nucleotides can be regarded as the building blocks for the larger nucleic acid molecules, DNA and RNA.

NUCLEOSIDE DI- AND TRIPHOSPHATES

One or two additional phosphates can be added to the first phosphate group of a nucleoside molecule (a nucleoside monophosphate, NMP) by means of a pyrophosphate linkage. The molecules formed in this way are called nucleoside diphosphates (NDPs) and nucleoside triphosphates (NTPs). The most important base involved in these compounds is adenine, forming the adenosine mono-, di- and triphosphate molecules (AMP, ADP and ATP), which fulfil vital roles in many cellular processes as you may know from your study of respiration and photosynthesis [for further details see BASC Booklets 4 (Metabolism) and 6 (Photosynthesis)].

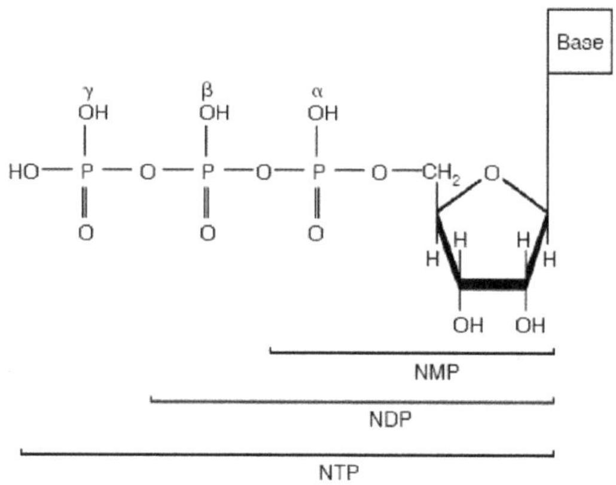

Fig. 7

NUCLEIC ACIDS

Nucleic acids are formed by the combination of nucleotide molecules through sugar–phosphate bonds known as phosphodiester linkages. Because a nucleic acid is a polymer of many nucleotide molecules, DNA and RNA molecules are called polynucleotides (Fig. 8).

The structure of a polynucleotide is shown diagrammatically above. In common with the formation of other biological polymers, e.g. starch and proteins, water molecules are produced by condensation reactions and energy from ATP molecules, produced in cellular respiration, is used.

From Fig. 9, it can be seen that the phosphodiester bond is formed between the 5carbon of one sugar molecule and the 3 carbon of the next.

The double-stranded nature of DNA is shown schematically above. Since it would be very time consuming to have to write out such a structure every time it was needed, abbreviated forms have been introduced. These are essentially of three types:

If the 'P' is on the left of a base then it is attached to the 5-carbon (five); if it is on the right then it is attached to the 3-carbon (three).

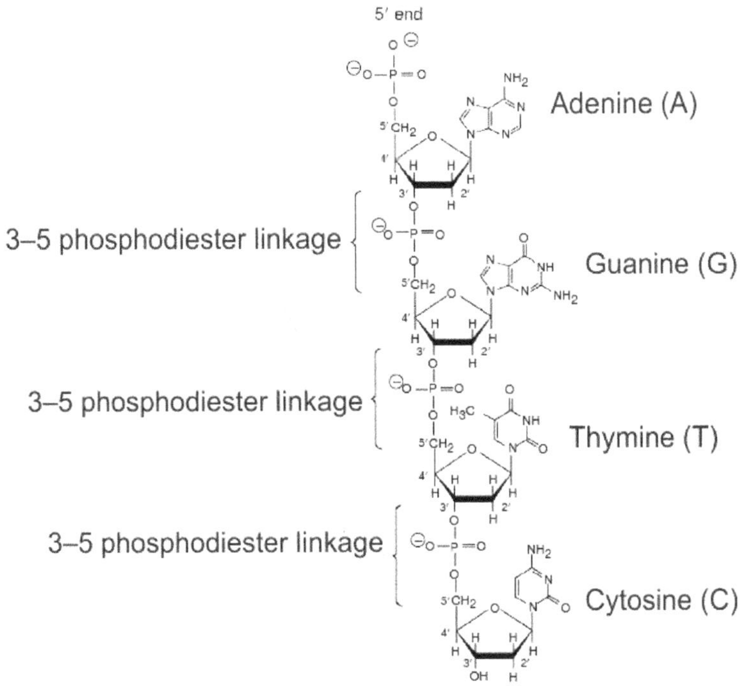

Fig 8

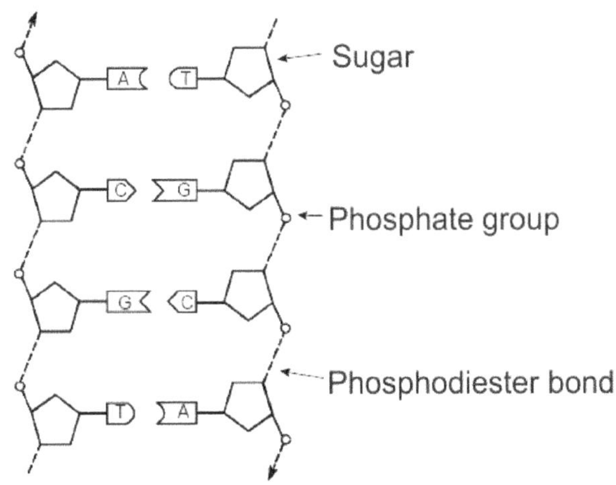

Fig 9

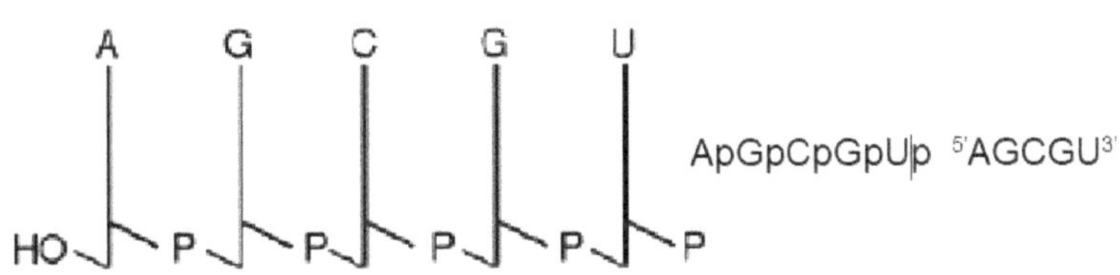

Fig. 10

DNA ISOLATION,BASE COMPOSITION AND STRUCTURE

Isolation

In order to study the chemistry of DNA it must be isolated and purified from the nucleus of a cell. In the isolation of DNA from cell samples, the major contaminant is histone protein which is closely associated with DNA to form visible chromatin threads in the nucleus. To remove this contaminant the cell is first deproteinized by treatment with phenol and the DNA is precipitated using ice-cold ethanol. A more detailed account of the isolation of DNA is given in one of the practical schedules later and in BASC Booklet 7, Recombinant DNA Technology.

Once isolated, DNA can be broken down or degraded by either: (i) treatment with perchloric acid at 100C, which breaks down the DNA to its constituent bases; or (ii) treatment with enzymes (DNase, phosphodiesterase, etc.) which break down the DNA to its constituent nucleotides.

The mixture of either of these products can be separated by chromatography or electrophoresis and the amount of each of the constituent bases can be measured.

Composition

Using the techniques outlined above it was noted by Chargaff and coworkers in the late 1940s that:

i. the base composition of an organism's DNA is characteristic of that organism;
ii. different cells or tissues of the same organism have identical base composition;
iii. closely related organisms exhibit similar base compositions (this property is used in biology as a basis for chemical taxonomy)
iv. the majority of DNA molecules exhibit certain chemical regularities that are defined as Chargaff's base pairing rules (base equivalence), namely the amount of A equals the amount of T and the amount of G equals the amount of C.

In considering the data shown in Table 2, in terms of base equivalence it is necessary to combine the values for cytosine and 5-methyl cytosine where these apply. The only exception to base equivalence is the DNA from the virus X174. Can you think of an explanation?

Structure

In 1947 Astbury studied DNA using the technique of X-ray diffraction, shown schematically below.

When the beam of X-rays collides with an electron-dense region in the DNA, the beam is diffracted. From the pattern of the diffracted beam it is possible to determine the arrangement of atoms in the DNA. In this way, Astbury was able to deduce that there was a regular feature occurring within the DNA molecule every 3.34 Å. Such a regular feature is referred to as a crystallographic repeat or identity period.

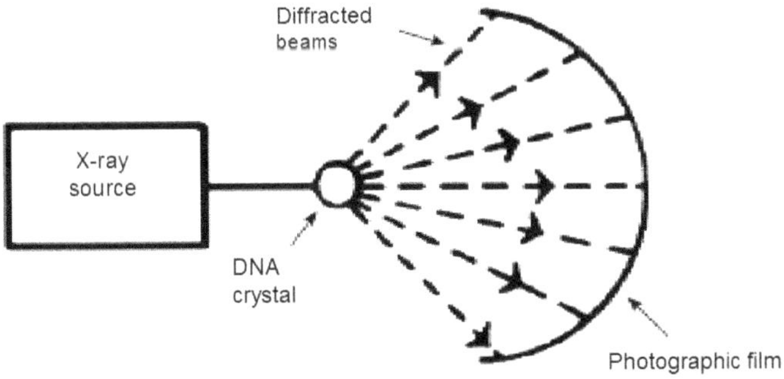

Fig. 11

Later work by Franklin, Goslin, Wilkins and co-workers (1953) demonstrated that there were two forms of DNA, a crystalline form known as the A form and a paracrystalline form known as the B form.

Based on the X-ray diffraction work and Chargaff's base equivalence rules, Watson and Crick in 1953 deduced the structure of DNA as a double helix. This comprised two strands of sugar–phosphate backbone with the bases positioned in between, with adenine (A) on one strand being hydrogen (or H-) bonded to a thymine (T) on the other, and a guanine (G) on one strand being H-bonded to a cytosine (C) on the other. The H-bonding arrangement of these base pairs is shown below (Fig. 12)

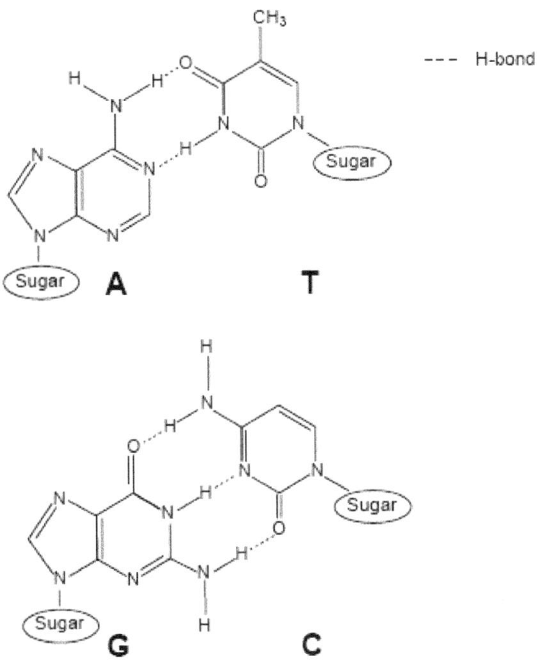

Fig. 12

A single-ringed pyrimidine base opposite a double-ringed purine base would ensure equal distance between the two polynucleotide strands but why A–T and G–C specifically? The answer lies in the number of H-bonds that can be formed between the bases as shown.

Because the two polynucleotide strands have 'direction', two configurations are possible, one in which both strands run in the same direction, which is known as a parallel arrangement, and one in which the strands run in the opposite direction, which is known as an antiparallel arrangement.

It has been shown that the DNA duplexes found in nature have their chains in an antiparallel arrangement. This has a profound significance when we come to consider the biological function of the molecules. In coiling the backbone into a helical configuration, two arrangements in space are possible, one a right-handed helix, the other a lefthanded helix. The difference between these two arrangements can be easily demonstrated using two pipe cleaners. It has been shown that the double helix of DNA found in nature is in the right-handed configuration.

Subsequent studies using space-filling models showed that there were two grooves in the DNA molecule, one shallow (approx. 12 Å) and one deep (approx. 22 Å). This feature was found to be important in relation to the binding of certain antibiotics, polymerase enzymes, etc. The principal features of the structure of DNA are summarized in the schematic drawing shown below.

Earlier in this chapter it was observed that the DNA isolated from the virus X174 did not exhibit base equivalence. The reason for this is that the DNA of this virus is single-stranded and not the normal duplex form. This has a significance for its mode of replication.

RNA ISOLATION, BASE COMPOSITION AND STRUCTURE ISOLATION

The isolation procedure is very similar to that for DNA isolation although it is essential to include an RNase inhibitor to prevent degradation of the RNA. Unlike DNA, which exists as a more or less homogeneous molecule for any one cell, RNA exists as a family of molecules, each member of which has a distinctive structure and function (see later sections in this chapter). Because of this wide variety of RNA species it is necessary to separate further the initial RNA pool into its constituent molecules. The methods employed include:

- Density-gradient centrifugation;
- Ion-exchange chromatography;
- Gel filtration;
- Electrophoresis.

Base composition

Purified RNA molecules can be degraded by chemical means (alkali) or by enzymes (RNase) and the base composition determined in a manner similar to that described for DNA. One major difference between DNA and RNA is that in RNA there are a fairly large number of minor bases. Also, in the majority of cases there is no base equivalence, signifying that RNA molecules are usually single-stranded.

Structure

There are three major forms of RNA: ribosomal RNA (rRNA), transfer RNA (tRNA) and messenger RNA (mRNA) and these exist in all the major life forms, together with other distinct RNA molecules which are not universal. rRNA accounts for about 80% of the total cellular RNA and is associated with protein to form the cytoplasmic particles known as ribosomes. As shown below the ribosome itself can be considered as two subunits, a large and a small, both of which are RNA–protein associations with about 65% RNA: 35% protein. The RNA has high molecular mass and is metabolically stable.

Transfer RNA is the next most abundant species and accounts for about 15% of the total RNA. These molecules are of much lower molecular mass than the rRNA and are also referred to as 4S RNA. As we shall see later, these molecules function as adaptors for amino acids in the course of protein synthesis and many different tRNAs exist, each being specific for one amino acid. The tRNA molecule is single-stranded but the chain folds back on itself in a very distinctive way (Fig. 13) such that about 50–60% of the structure is base-paired.

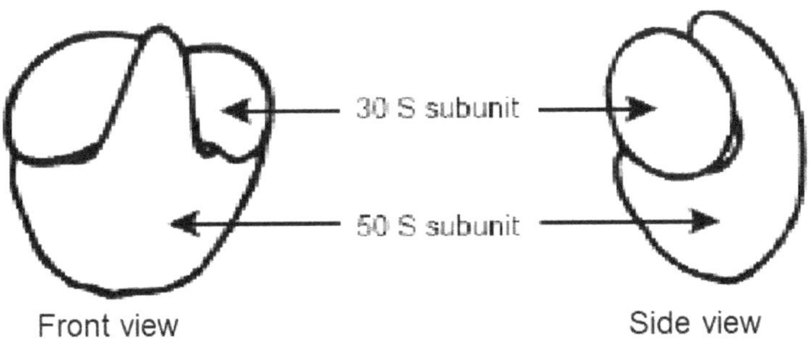

Fig. 13

Most of the remainder of the cell RNA (less than 5%) is accounted for by mRNA which, in eukaryotes, originates in the nucleus and migrates to the cytoplasm during protein synthesis. It is of high molecular mass and is metabolically very labile, i.e. easily broken down. As we shall see in a later section, mRNA is centrally involved in the transfer and expression of the genetic information and is responsible for the sequence of amino acids in each of the different proteins in the cell. Because the size of the messenger is variable, no S-value is associated with it.

SYNTHESIS OF NUCLEIC ACIDS

Introduction

Advances in DNA synthesis, which have made it possible to widely use manmade DNAs in molecular biology, genetics, virology, and other fields, represent the most important contribution to the chemistry of nucleic acids in recent years. Along with achievements in molecular cloning and nucleic acid sequencing, DNA synthesis opens up broad possibilities for future progress in biology. Synthetic DNAs and their fragments are already extensively employed in the synthesis and isolation of genes, in protein engineering, as probes in the diagnosis of infectious and genetic diseases, and, last but not least, as gene-targeted agents.

The availability of synthetic DNA fragments with a predetermined monomer sequence, which has made the above advances in molecular biology, genetic engineering and other related fields possible, was ensured by automation of their chemical synthesis and developments in purification of the preparations. The entire process of synthesizing gene fragments, assembling them into genes, cloning, selection of clones and structural analysis is now a matter of just a few weeks. Armed with powerful tools, the organic chemist is entering a new era. It has become possible to study the structure-activity relationship in proteins as well as to shape their three-dimensional structures in the course of DNA-dependent synthesis in vitro. According to predictions, new enzymes with predetermined functions will be designed and synthesized by the turn of the century.

Emphasis in this chapter will be placed on the chemical principles underlying synthesis of oligodeoxyribonucleotides.

Currently, synthesis of oligonucleotides in automatic synthesizers is becoming routine even in laboratories without specialists in synthetic organic chemistry. This places special importance on reviews and textbooks summarizing the latest knowledge about DNA synthesis not only for the benefit of students but also scientists involved in synthetic oligo(poly)nueleotide research.

Chemical synthesis of oligo- and polynucleotides represents the most exciting and challenging field of synthetic chemistry of nucleic acids. In establishing an internucleotide bond (chemical linkage of monomer units), the problem boils down to using the phosphate of one monomer unit (nucleotide component) to acylate the hydroxyl group of another unit (nucleoside component):

Enzymatic linking of monomer units involves a "directed" reaction or, to be more precise, activation of the phosphate and formation of a natural 3'-5'internucleotide bond. The enzyme "takes care" of directed activation and linkage of groups capable of intermonomer bonding. To achieve the same end by chemical synthesis requires accomplishment of at least two tasks:

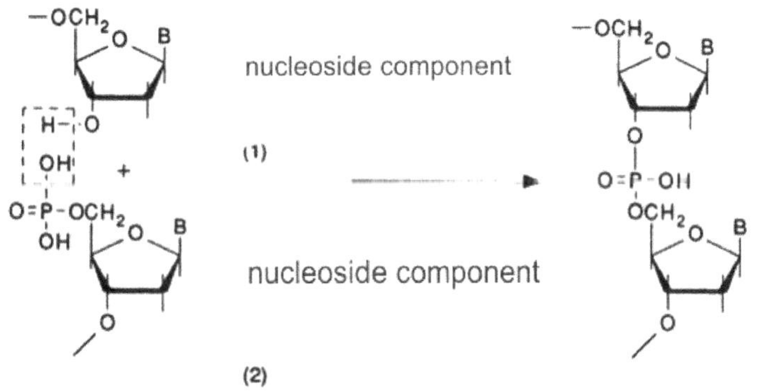

Fig 14

i. Protection of all functional groups of the monomer, not subject to phosphorylation;
ii. Activation of the phosphate group in the nucleotide or, in other words, transforming the nucleotide to a phosphorylating agent.

If these tasks are compared with the similar ones arising during peptide bond synthesis, it becomes evident that the synthesis of oligonucleotides gives rise to many more difficulties. There are basically two reasons for that:

(a) a great number of functional groups in the monomer units of nucleic acids and presence of sufficiently labile bonds in the latter; and

(b) much lower activity of the reacting groups (the hydroxyls of ribose make a much weaker nucleophile in comparison with the amino group of amino acids, whereas the phosphate group in nucleotides is a weaker electrophile than the carbonyl of amino acids).

As a result, the preparation of monomers - nucleoside (1) and nucleotide (2) components for directed synthesis of the 3'-5'-internucleotide bond - is extremely difficult. The highly selective blocking and deblocking of the hydroxy groups in pentose, the amino groups of the heterocycles, and the phosphate groups in the nucleoside component (which must remain intact) should be carried out under very mild conditions; at the same time, the phosphate group must be active enough to acylate the desired hydroxyl of the sugar. Another difficulty in oligonucleotide synthesis stems from the presence of two hydroxyls in the nucleotide's phosphate group.

Once an oligonucleotide has been synthesized, the internucleotide phosphate does not remain neutral. It may react, for example, with the activated phosphate present in the reaction mixture, which significantly affects the oligonucleotide yields. To circumvent this difficulty, methods have been developed to ensure internucleotide linkage not of the phosphodiester type (3) (phosphodiester method of synthesis) but of the phosphotriester type (4) (phosphotriester method).

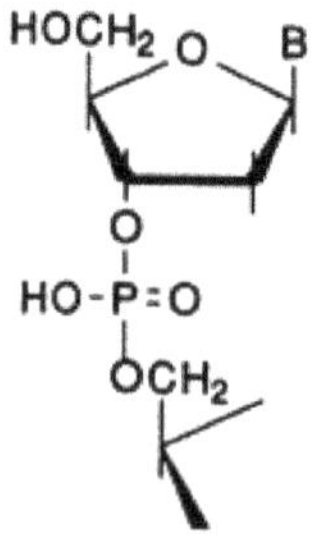

(3)

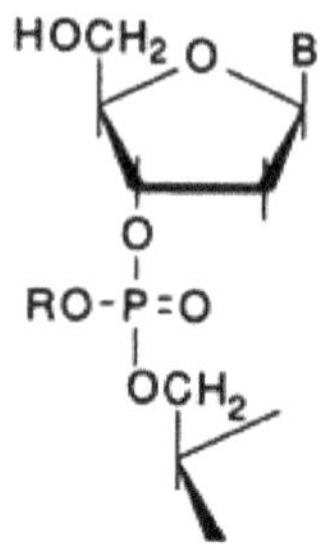

(4)

Fig. 15

To create natural 3’-5’-internucleotide bonds one can, in principle, resort to phosphorylation of the corresponding nucleoside components with both 5’phosphates (path I) and 3’-phosphates (path II):

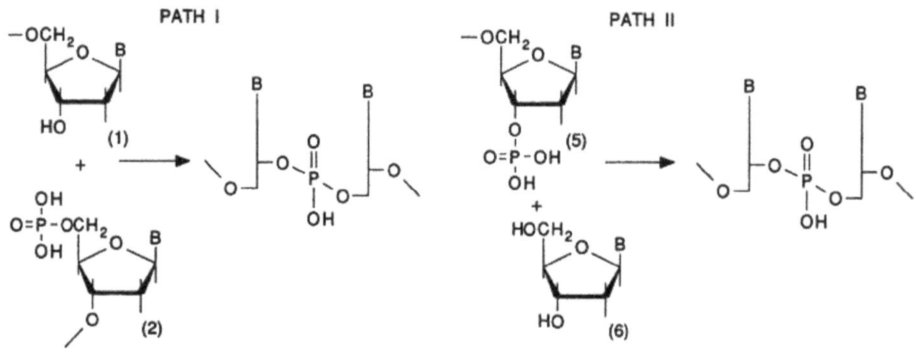

Fig. 16

21

STRUCTURE AND SYNTHESIS OF PROTEINS

PROTEIN STRUCTURE

Proteins are the end products of the decoding process that starts with the information in cellular DNA. As workhorses of the cell, proteins compose structural and motor elements in the cell, and they serve as the catalysts for virtually every biochemical reaction that occurs in living things. This incredible array of functions derives from a startlingly simple code that specifies a hugely diverse set of structures.

In fact, each gene in cellular DNA contains the code for a unique protein structure. Not only are these proteins assembled with different amino acid sequences, but they also are held together by different bonds and folded into a variety of three-dimensional structures. The folded shape, or conformation, depends directly on the linear amino acid sequence of the protein.

The building blocks of proteins are amino acids, which are small organic molecules that consist of an alpha (central) carbon atom linked to an amino group, a carboxyl group, a hydrogen atom, and a variable component called a side chain. Within a protein, multiple amino acids are linked together by peptide bonds, thereby forming a long chain. Peptide bonds are formed by a biochemical reaction that extracts a water molecule as it joins the amino group of one amino acid to the carboxyl group of a neighboring amino acid. The linear sequence of amino acids within a protein is considered the primary structure of the protein.

Proteins are built from a set of only twenty amino acids, each of which has a unique side chain. The side chains of amino acids have different chemistries. The largest group of amino acids have nonpolar side chains. Several other amino acids have side chains with positive or negative charges, while others have polar but uncharged side chains. The chemistry of amino acid side chains is critical to protein structure because these side chains can bond with one another to hold a length of protein in a certain shape or conformation. Charged amino acid side chains can form ionic bonds, and polar amino acids are capable of forming hydrogen bonds. Hydrophobic side chains interact with each other via weak van der Waals interactions. The vast majority of bonds formed by these side chains are noncovalent. In fact, cysteines are the only amino acids capable of forming covalent bonds, which they do with their particular side chains. Because of side chain interactions, the sequence and location of amino acids in a particular protein guides where the bends and folds occur in that protein (Fig. 1).

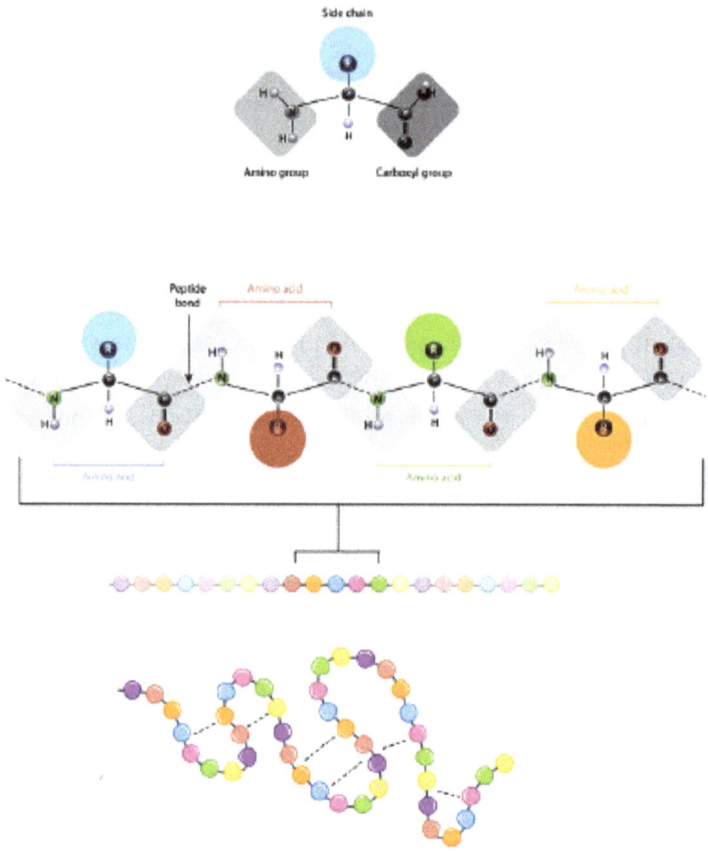

Fig. 1:The relationship between amino acid side chains and protein conformation

The primary structure of a protein — its amino acid sequence — drives the folding and intramolecular bonding of the linear amino acid chain, which ultimately determines the protein's unique three-dimensional shape. Hydrogen bonding between amino groups and carboxyl groups in neighboring regions of the protein chain sometimes causes certain patterns of folding to occur. Known as alpha helices and beta sheets, these stable folding patterns make up the secondary structure of a protein. Most proteins contain multiple helices and sheets, in addition to other less common patterns (Fig. 2). The ensemble of formations and folds in a single linear chain of amino acids — sometimes called a polypeptide — constitutes the tertiary structure of a protein. Finally, the quaternary structure of a protein refers to those macromolecules with multiple polypeptide chains or subunits.

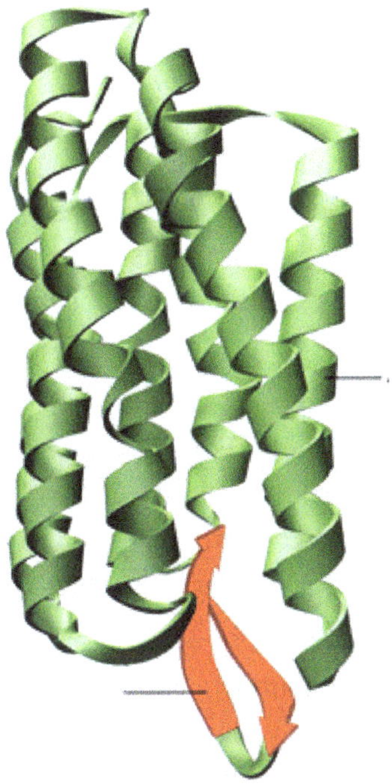

Fig. 2: The structure of the protein bacteriorhodopsin

The final shape adopted by a newly synthesized protein is typically the most energetically favorable one. As proteins fold, they test a variety of conformations before reaching their final form, which is unique and compact. Folded proteins are stabilized by thousands of noncovalent bonds between amino acids. In addition, chemical forces between a protein and its immediate environment contribute to protein shape and stability.

For example, the proteins that are dissolved in the cell cytoplasm have hydrophilic (water-loving) chemical groups on their surfaces, whereas their hydrophobic (water-averse) elements tend to be tucked inside. In contrast, the proteins that are inserted into the cell membranes display some hydrophobic chemical groups on their surface, specifically in those regions where the protein surface is exposed to membrane lipids. It is important to note, however, that fully folded proteins are not frozen into shape. Rather, the atoms within these proteins remain capable of making small movements.

Even though proteins are considered macromolecules, they are too small to visualize, even with a microscope. So, scientists must use indirect methods to figure out what they look like and how they are folded. The most common method used to study protein structures is X-ray crystallography.

With this method, solid crystals of purified protein are placed in an X-ray beam, and the pattern of deflected X rays is used to predict the positions of the thousands of atoms within the protein crystal.

How Do Proteins Arrive at Their Final Shapes?

In theory, once their constituent amino acids are strung together, proteins attain their final shapes without any energy input. In reality, however, the cytoplasm is a crowded place, filled with many other macromolecules capable of interacting with a partially folded protein. Inappropriate associations with nearby proteins can interfere with proper folding and cause large aggregates of proteins to form in cells. Cells therefore rely on so-called chaperone proteins to prevent these inappropriate associations with unintended folding partners.

Chaperone proteins surround a protein during the folding process, sequestering the protein until folding is complete. For example, in bacteria, multiple molecules of the chaperone GroEL form a hollow chamber around proteins that are in the process of folding. Molecules of a second chaperone, GroES, then form a lid over the chamber. Eukaryotes use different families of chaperone proteins, although they function in similar ways.

Chaperone proteins are abundant in cells. These chaperones use energy from ATP to bind and release polypeptides as they go through the folding process. Chaperones also assist in the refolding of proteins in cells. Folded proteins are actually fragile structures, which can easily denature, or unfold. Although many thousands of bonds hold proteins together, most of the bonds are noncovalent and fairly weak. Even under normal circumstances, a portion of all cellular proteins are unfolded. Increasing body temperature by only a few degrees can significantly increase the rate of unfolding. When this happens, repairing existing proteins using chaperones is much more efficient than synthesizing new ones. Interestingly, cells synthesize additional chaperone proteins in response to "heat shock."

The Protein Families

All proteins bind to other molecules in order to complete their tasks, and the precise function of a protein depends on the way its exposed surfaces interact with those molecules. Proteins with related shapes tend to interact with certain molecules in similar ways, and these proteins are therefore considered a protein family. The proteins within a particular family tend to perform similar functions within the cell.

Proteins from the same family also often have long stretches of similar amino acid sequences within their primary structure. These stretches have been conserved through evolution and are vital to the catalytic function of the protein. For example, cell receptor proteins contain different amino acid sequences at their binding sites, which receive chemical signals from outside the cell, but they are more similar in amino acid sequences that interact with common intracellular signaling proteins. Protein families may have many members, and they likely evolved from ancient gene duplications. These duplications led to modifications of protein functions and expanded the functional repertoire of organisms over time.

PROTEIN SYNTHESIS

This amazing artwork (Figure 5.7.1) shows a process that takes place in the cells of all living things: the production of proteinsno post. This process is called protein synthesis, and it actually consists of two processes — transcription and translation. In eukaryotic cells, transcription takes place in the nucleus. During transcription, DNA is used as a template to make a molecule of messenger RNA (mRNA). The molecule of mRNA then leaves the nucleus and goes to a ribosome in the cytoplasm, where translation occurs. During translation, the genetic code in mRNA is read and used to make a polypeptide. These two processes are summed up by the central dogma of molecular biology: DNA → RNA → Protein.

Transcription

Transcription is the first part of the central dogma of molecular biology: DNA → RNA. It is the transfer of genetic instructions in DNA to mRNA. During transcription, a strand of mRNA is made to complement a strand of DNA. You can see how this happens in Fig. 3.

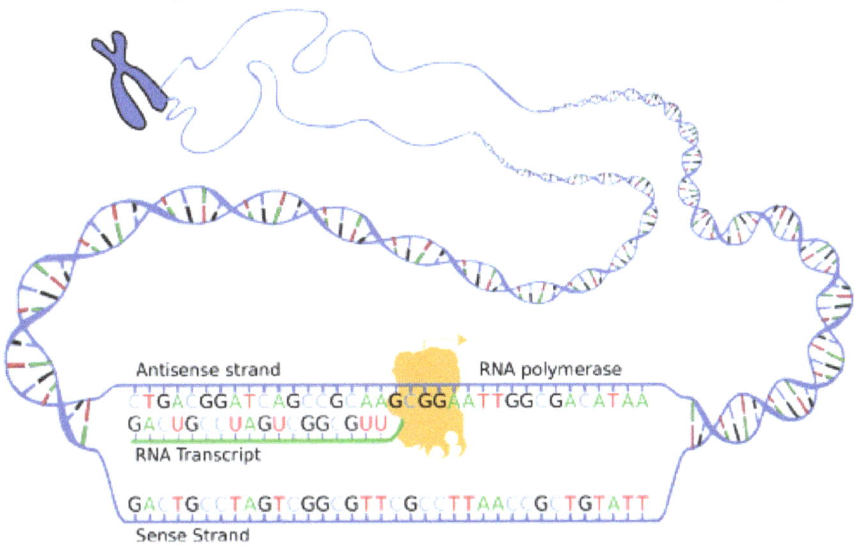

Fig 3 Transcription uses the sequence of bases in a strand of DNA to make a complementary strand of mRNA. Triplets are groups of three successive nucleotide bases in DNA. Codons are complementary groups of bases in mRNA.

Transcription begins when the enzyme RNA polymerase binds to a region of a gene called the promoter sequence. This signals the DNA to unwind so the enzyme can "read" the bases of DNA. The two strands of DNA are named based on whether they will be used as a template for RNA or not. The strand that is used as a template is called the template strand, or can also be called the antisense strand. The sequence of bases on the opposite strand of DNA is called the non-coding or sense strand. Once the DNA has opened, and RNA polymerase has attached, the RNA polymerase moves along the DNA, adding

RNA nucleotides to the growing mRNA strand. The template strand of DNA is used as to create mRNA through complementary base pairing. Once the mRNA strand is complete, and it detaches from DNA. The result is a strand of mRNA that is nearly identical to the coding strand DNA – the only difference being that DNA uses the base thymine, and the mRNA uses uracil in the place of thymine

Processing mRNA

In eukaryotes, the new mRNA is not yet ready for translation. At this stage, it is called pre-mRNA, and it must go through more processing before it leaves the nucleus as mature mRNA. The processing may include splicing, editing, and polyadenylation. These processes modify the mRNA in various ways. Such modifications allow a single gene to be used to make more than one protein.

- Splicing removes introns from mRNA, as shown in Figure 4. Introns are regions that do not code for the protein. The remaining mRNA consists only of regions called exons that do code for the protein. The ribonucleoproteins in the diagram are small proteins in the nucleus that contain RNA and are needed for the splicing process.

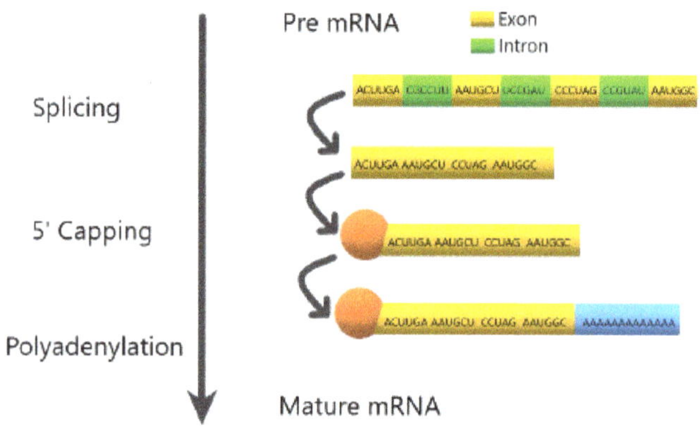

Fig. 4 Pre mRNA processing. mRNA requires processing before it leaves the nucleus.

- Editing changes some of the nucleotides in mRNA. For example, a human protein called APOB, which helps transport lipids in the blood, has two different forms because of editing. One form is smaller than the other because editing adds an earlier stop signal in mRNA.
- 52 Capping adds a methylated cap to the "head" of the mRNA. This cap protects the mRNA from breaking down, and helps the ribosomes know where to bind to the mRNA
- Polyadenylation adds a "tail" to the mRNA. The tail consists of a string of As (adenine bases). It signals the end of mRNA. It is also involved in exporting mRNA from the nucleus, and it protects mRNA from enzymes that might break it down.

Translation

Translation is the second part of the central dogma of molecular biology: RNA → Protein. It is the process in which the genetic code in mRNA is read to make a protein. Translation is illustrated in Fig. 5. After mRNA leaves the nucleus, it moves to a ribosome, which consists of rRNA and proteins. The ribosome reads the sequence of codons in mRNA, and molecules of tRNA bring amino acids to the ribosome in the correct sequence.

Translation occurs in three stages: Initiation, Elongation and Termination.

Initiation

After transcription in the nucleus, the mRNA exits through a nuclear pore and enters the cytoplasm. At the region on the mRNA containing the methylated cap and the start codon, the small and large subunits of the ribosome bind to the mRNA. These are then joined by a tRNA which contains the anticodons matching the start codon on the mRNA. This group of molecues (mRNA, ribosome, tRNA) is called an initiation complex.

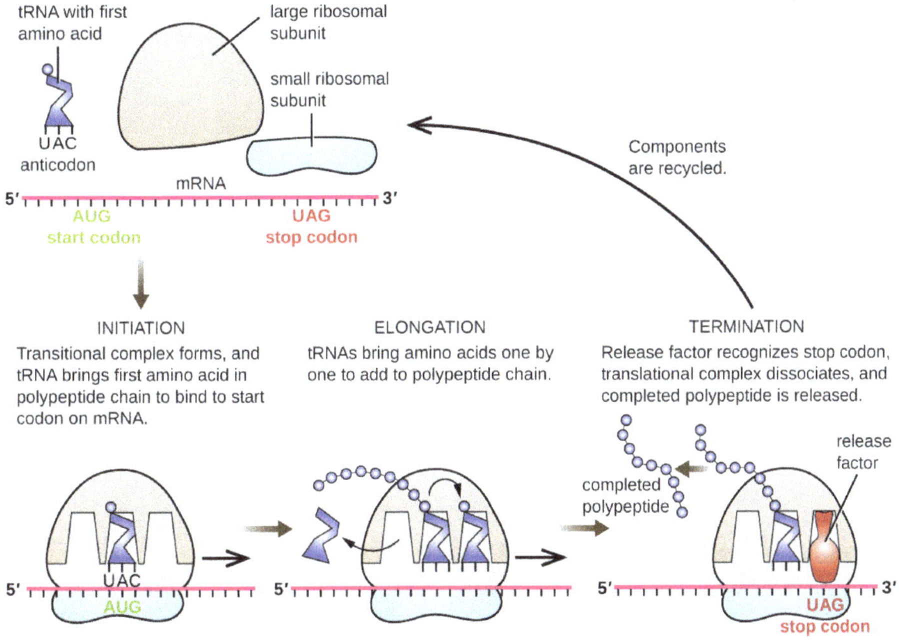

Fig. 5 Translation takes place in three stages: Initiation, Elongation and Termination.

Elongation

tRNA keep bringing amino acids to the growing polypeptide according to complementary base pairing between the codons on the mRNA and the anticodons on the tRNA. As a tRNA moves into the ribosome, its amino acid is transferred to the growing polypeptide. Once this transfer is complete, the tRNA leaves the ribosome, the ribosome moves one codon length down the mRNA, and a new tRNA enters with its corresponding amino acid. This process repeats and the polypeptide grows.

Termination

At the end of the mRNA coding is a stop codon which will end the elongation stage. The stop codon doesn't call for a tRNA, but instead for a type of protein called a release factor, which will cause the entire complex (mRNA, ribosome, tRNA, and polypeptide) to break apart, releasing all of the components.

SUMMARY OF PROTEIN STRUCTURE AND SYNTHESIS

There are hierarchical degrees of complexity in the formation of a mature, fully-folded protein. Primary structure describes a linear chain of amino acids formed by sequential peptide bonds. Secondary structure consists of backbone interactions in the polypeptide chain that are stabilized by hydrogen bonds between amino acids, forming α-helices and β-pleated sheets. In tertiary structure, secondary structures come together and further orient themselves depending on hydrophobic interactions, ionic interactions, and disulfide bridges. Finally, quaternary structure consists of multiple tertiary structures coming together to form subunits of a larger protein.

22

REGULATION OF PROTEIN SYNTHESIS

PROTEIN SYNTHESIS

The central dogma describes how the DNA makes its own copies by DNA replication, then codes for the RNA by means of a transcription and RNA codes for the proteins by translation.

For the creation of a new generation, the genetic information stored in the DNA has to be transferred from one generation to another. The process begins with the replication of DNA followed by transcription and translation. Let us have a discussion on translation viz. protein synthesis and the structure of tRNA.

STRUCTURE OF TRNA

The DNA is transcribed into mRNA on the basis of their complementarity. Translation is the process of conversion of nucleic acid information into amino acids. There is no complementarity between amino acids and mRNA. Hence, translation is not controlled by complementarity but by the genetic code. Since amino acids cannot read this genetic code, they are dependent on an adapter molecule. This adapter molecule is called tRNA (transfer RNA).

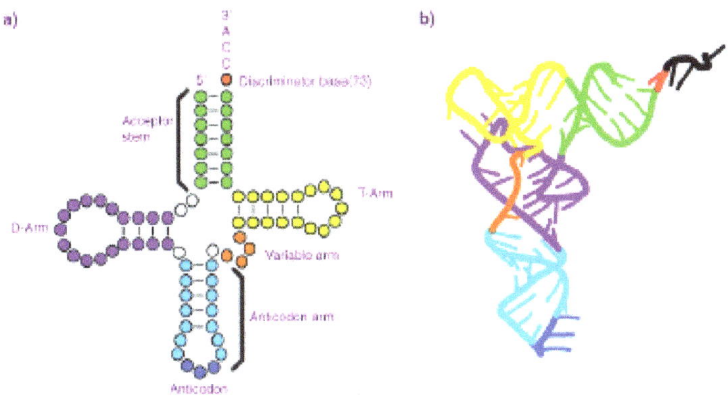

Fig. 1. Structure of tRNA

Structurally, the tRNA is an inverted L-shaped molecule which has an anticodon loop and amino acid acceptor end. The anticodon loop makes bases complementary to the codes on the mRNA and amino acid end binds to the respective amino acids. Thus, it helps in protein synthesis. Each amino acid has a specific tRNA. Initiator tRNA initiates the translation while stop codons have no tRNA.

PROTEIN SYNTHESIS - TRANSLATION

The process by which the mRNA codes for a particular protein is known as Translation. In the process, the ribosome translates the mRNA produced from DNA into a chain of specific amino acids. This chain of amino acids leads to protein synthesis. It is a process where the expense of ATP is required and this energy is given by the charged tRNA. The whole machinery of translation is present in the ribosomes.

The ribosomes consist of a bigger subunit and a smaller subunit. The larger subunit, in turn, consists of two molecules of tRNA that are placed close so that peptide bond could be developed at the expense of sufficient energy. The mRNA enters the smaller subunit which is held by the molecules of tRNA of the complementary codon, that exists in the bigger subunit. Hence, two codons are held by two molecules of tRNA, placed near to each other and a peptide bond is produced among them. When this process repeats, a large chain of amino acids is synthesized. At the stop, codon ribosome releases the amino acid chain.

REGULATION OF TRANSLATION

We already saw how miRNAs can inhibit translation, but there are a number of other ways that translation of an mRNA can also be regulated in a cell. One key step for regulation is translation initiation.

In order for translation to begin, the ribosome, an RNA-and-protein complex that houses translation, must assemble on the mRNA. This process involves many "helper" proteins, which make sure the ribosome is correctly positioned. Translation can be regulated globally (for every mRNA in the cell) through changes in the availability or activity of the "helper" proteins.

For example, in order for translation to begin, a protein called eukaryotic initiation factor-2 (eIF-2) must bind to a part of the ribosome called the small subunit. Binding of eIF-2 is controlled by phosphorylation, or addition of a phosphate group to the protein.

When eIF-2 is phosphorylated, it's turned "off"—it undergoes a shape-change and can no longer play its role in initiation, so translation cannot begin. When eIF-2 is not phosphorylated, in contrast, it's "on" and can carry out its role in initiation, allowing translation to proceed.

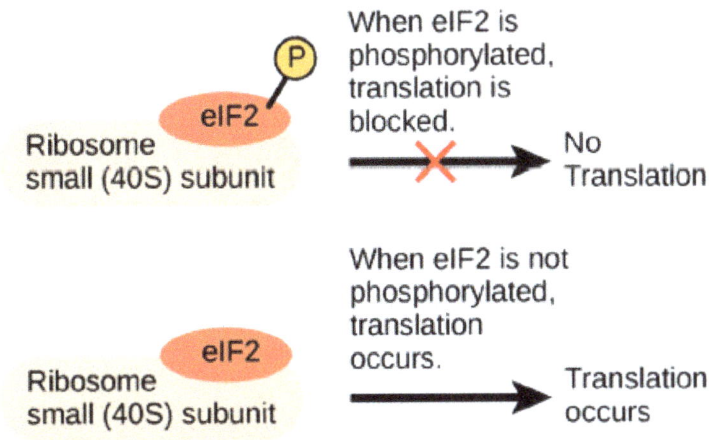

Fig. 2. Gene expression can be controlled by factors that bind the translation initiation complex.

CHEMICAL MODIFICATIONS, PROTEIN ACTIVITY, AND LONGEVITY

Proteins can be chemically modified with the addition of groups including methyl, phosphate, acetyl, and ubiquitin groups. The addition or removal of these groups from proteins regulates their activity or the length of time they exist in the cell. Sometimes these modifications can regulate where a protein is found in the cell—for example, in the nucleus, in the cytoplasm, or attached to the plasma membrane.

Chemical modifications occur in response to external stimuli such as stress, the lack of nutrients, heat, or ultraviolet light exposure. These changes can alter epigenetic accessibility, transcription, mRNA stability, or translation—all resulting in changes in the expression of various genes. This is an efficient way for the cell to rapidly change the levels of specific proteins in response to the environment. Because proteins are involved in every stage of gene regulation, the phosphorylation of a protein (depending on the protein that is modified) can: alter accessibility to the chromosome; alter translation (by altering transcription factor binding or function); change nuclear shuttling (by influencing modifications to the nuclear pore complex); alter RNA stability (by binding or not binding to the RNA to regulate its stability); modify translation (increase or decrease); or change post-translational modifications (add or remove phosphates or other chemical modifications).

The addition of a ubiquitin group to a protein marks that protein for degradation. Ubiquitin acts like a flag indicating that the protein lifespan is complete. These proteins are moved to the proteasome, an organelle that functions to remove proteins, to be degraded. One way to control gene expression, therefore, is to alter the longevity of the protein.

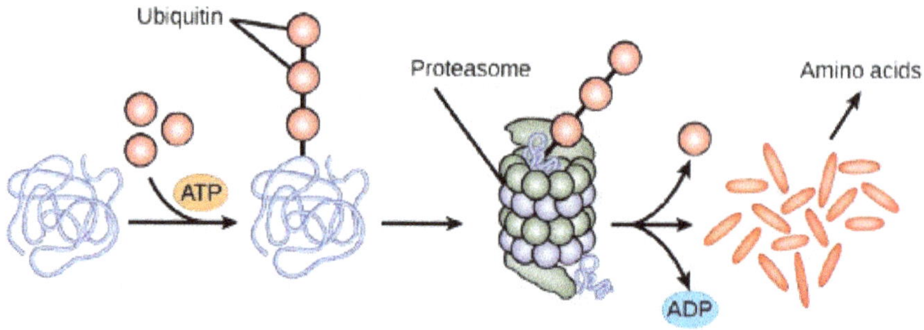

Proteins with ubiquitin tags are marked for degradation within the proteasome.

23

GENETIC CODE, OVERLAPING GENES AND SPLIT GENES

A split gene consists of both exons and introns when the post-transcriptional modification happens, the introns which are non-coding genes are removed and only exons which are protein coding genes are retained in the mature RNA.

Generally, the different genes occur as discrete units separate from each other, but some prokaryotic genes may be overlapping. Virtually all the prokaryotic genes are functional units. But eukaryotic genes present some interesting organisations such as split genes, overlapping genes, pseudogenes, etc.

It has been generally believed that a gene is a continuous, uninterrupted sequence of nucleotides which codes for a single polypeptide chain. Some recent work has shown that this may not always be so. The sequences of some eukaryotic genes (globin, ovalbumin) are found to be interrupted by nucleotides that are not represented within the amino acid sequence of the protein.

Although these sequences are transcribed into hnRNA, they are later excised (spliced) and removed, so that they are not included in the mature mRNA that is translated into protein (Fig. 1). Such interruptions within the sequence of a gene have been variously called as introns, inserts, intervening sequences or 'silent' DNA by some.

The sequences which are included in the mRNA and translated have been called exons; the eukaryotic genes being a mosaic of introns and exons. Although the coding regions are interrupted, they are present in the same order in the genome as in the mRNA. Hence the name split genes.

Intervening sequences were unequivocally demonstrated for the first time in the gene for β-globin in mouse and rabbit and then in chick ovalbumin. In the β-globin gene of mouse and rabbit a single intervening sequence is present about 200-250 nucleotides inside from one end of the gene, in both somatic and germ line cells of the animal.

White and Hogness (1977) have developed an R loop mapping technique for identifying the site and length of an intervening sequence in the electron microscope. The method involves hybridisation of the RNA transcript with double helical DNA of the gene at high temperature and increased concentrations of formamide.

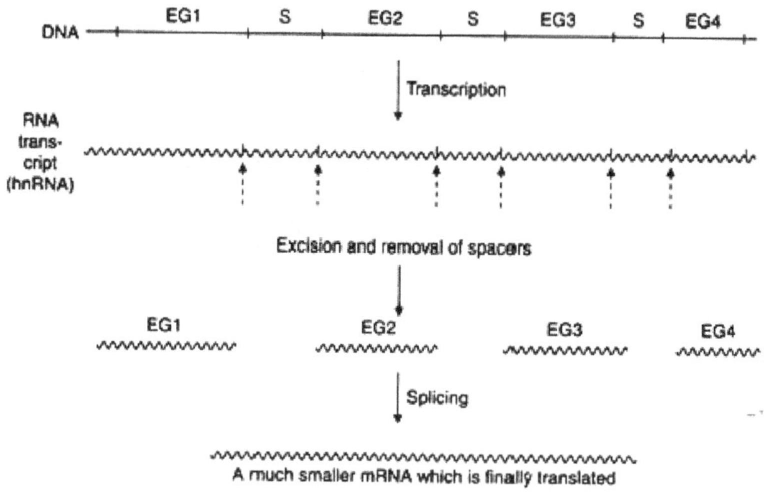

Fig. 1 Diagram to show relation between expressed genes (EG) and spacers(s) in split genes

SPLIT GENES OR INTERRUPTED GENES:

The coding regions containing actual information of the genes (exons) of most eukaryotic genes are interrupted by few to several noncoding sequences called introns (from intervening sequences) which are spliced out after transcription Such genes are called split genes since their coding sequences are split into several parts due to the introns. But some genes of eukaryotes are not split, e.g., histone genes of sea urchin.

The first split gene to be described in 1977 by Pierre Chambon and his colleagues was the ovalbumin genes of chicken coding for the 386 amino acid long ovalbumin protein of eggs.

Important features of interrupted genes

1. Each interrupted gene begins' with an exon and ends with an exon.
2. The exons occur in the same precise order in the mRNA in which they occur in the gene.
3. The same interrupted gene organisation is consistently present in all the tissues of organisms.
4. Most introns are blocked in all reading frames i.e., termination codons occur frequently in their three reading frames. Therefore, most introns do not seem to have coding functions.

Significance of Split Genes

The significance of split organisation of eukaryotic genes is not clear.

1. In some cases, different exons of a gene code for different active regions of the protein molecule, e.g., in the case of antibodies. Thus, it has been suggested that introns are relics of evolutionary processes that brought together different ancestral genes to form new larger genes. It is also possible that some introns have been introduced within certain exons during evolution.
2. Introns may also provide for increased recombination rates between exons of a gene and thus may be of some significance in genetic variation.
3. Introns are known to code for enzymes involved in the processing of hn RNA (heterogenous RNA).

OVERLAPPING GENES

Overlapping genes are defined as a pair of adjacent genes whose coding regions are partially overlapping. In other words, a single stretch of DNA codes for portions of two separate proteins. Such an arrangement of genetic code is ubiquitous. Many overlapping genes have been identified in the genomes of prokaryotes , eukaryotes , mitochondria , and viruses.

For two genes to overlap, the signal to begin transcription for one must reside inside the second gene, whose transcriptional start site is further "upstream." In addition, the "stop" signal for the second gene must not be read by the ribosome during translation , using the RNA copy of the gene. This is possible because RNA is read in triplets, meaning that it can contain three separate sequences that can be "read" by the cell's protein-making machinery. Such sequences of nucleotide triplets are called reading frames, and they are different in the RNA transcripts of the overlapping genes.

Overlapping genes enable the production of more proteins from a given region of DNA than is possible if the genes were arranged sequentially. Indeed, for the bacteriophage PhiX174, overlapping of genes is necessary. The amount of DNA present in the circular, single-stranded DNA genome of this virus would not be sufficient to encode the eleven bacteriophage proteins if transcription occurred in a linear fashion, one gene after another.

The genome economy afforded by overlapping genes extends to the human genome. The recently completed sequencing of the human genome has revealed between 30,000 and 70,000 genes. Yet evidence suggests that the human genome encodes 100,000 to 200,000 proteins. At least part of the information for the extra proteins may come from the presence of hitherto undiscovered overlapping genes, although more may come from alternative splicing of exons in a single gene.

In algae called *Guillardia*, a structure called a nucleomorph contains only about 500,000 base pairs of DNA, a very small genome, yet produces almost 500 proteins. Part of the efficient packaging of the genome is due to 44 overlapping genes. A nucleomorph is a remnant of a nucleus from an ancient eukaryotic organism that became incorporated into the algae.

One consequence of overlapping genes is to reduce the tolerance for mutation. Virtual experiments conducted within the past several years using a software system called Avida have indicated that overlapping reduces the probability of accumulating so-called neutral mutations in a gene (mutations that have no effect). Neutral mutations are unlikely with overlapping genes, because the mutation must have no effect on two genes with different reading frames.

The evolutionary origin of overlapping genes is not yet clear. Recent research indicates that they may have arisen due to several mutational events. These may include the loss of a signal to stop the transcription process in a gene and a shift in the reading sequence of the genetic components.

Pseudogenes

In multicellular organisms, a wide variety of DNA sequences are found, which are of no apparent use. Some of these sequences are defective copies of functional genes and are, therefore called pseudogenes. Their general organisation is similar to those of interrupted genes as they have sequences corresponding to introns and exons.

They are non-functional due to mutations which prevent one or more, often more of the following: transcription, RNA splicing and translation (due to frequent occurrence of termination codons). The pseudogenes are common features of gene clusters and occur with their functional counterparts. These pseudogenes have been reported in human beings, mouse and Drosophila. The most popular examples of these pseudogenes include the following:

i. Human α-globin and β-globin pseudogenes (ψ) found in each of the two globin clusters.
ii. In mouse there are few a-globin pseudogenes (ψ), one of them ($\psi\alpha_3$) is different from other pseudogenes since it has no introns which are present in a-globin genes as well as in other pseudogenes.
iii. UsnRNA series of pseudogenes in human beings includes U_1 U_2 and U_3sn RNA pseudogenes.
iv. In Drosophila histone pseudogenes have been discovered.

Several pseudogenes lack introns and resemble mature mRNA transcripts of their active counterparts, they are called processed pseudogenes. These pseudogenes are believed to be reverse transcripts of mature mRNAs which become inserted into the genome. Such genes are flanked by direct repeats of 6-21 bp and are located anywhere in the genome irrespective of the location of the concerned functional genes. For example, the mouse Ta 3 globin gene in mouse.

Gene Families

A gene family consists of all those genes that have related sequences and are believed to have originated form a common ancestral gene through gene duplication and subsequent mutational variation. The members of a gene family may be clustered together or dispersed on different chromosomes.

SPLIT GENES

This discovery provided one of the great surprises of 20th century molecular biology: genes of complex organisms are not present as continuous sections on DNA, but are interrupted by long stretches of non-coding DNA.

The coding parts are called exons, the non-coding sequences are called introns. The information of a gene is therefore fragmented and must be assembled during gene expression. In order to do this, an RNA copy of the gene is produced as a precursor: the pre-mRNA (precursor mRNA). The introns of this pre-mRNA are then recognized and cut out by a huge molecular machine, the spliceosome. This process is called splicing. After splicing and other maturation processes, the "mature" mRNA is present.

A split or interrupted gene is defined as a gene consisting of introns (intervening sequences between exons) and exons (segments of an interrupted gene that are represented in the mRNA). Thus, a simple split gene has at least two exons and one intron. Removal (splicing) of the intron(s) from a primary transcript (pre-mRNA) is an essential process for the creation of an mRNA. Thus, defining the exon–intron boundary is the first step to involve the accurate recognition of a 5' splice site (5' ss; the junction between the 5' end of the intron and the 3' end of the exon) and a 32' splice site (3' ss; the junction between the 3' end of the intron and the 5' end of the exon) by cellular splicing machinery. Most viral and eukaryotic RNA introns are GU-AG introns (see Fig. 2) presumably containing consensus sequences of 5' ss GURAGU and 3' ss YNYURAC-Py-AG (R, purine; Y, pyrimidine; N, any nucleotide; Py, polypyrimidine tract).

INTRON, EXON STRUCTURE OF SPLIT GENES:

Electron microscopic images of RNA-DNA hybrids and subsequent nucleotide sequencing of cloned genomic DNAs and complementary DNAs (cDNAs) demonstrated that the coding region of the mouse â-globin gene (encodes the â subunit of haemoglobin) is interrupted by two introns that are spliced out from mRNA.

In general, the intron-exon structure of eukaryotic genes is complicated, the amount of DNA in intron sequences usually exceeding that in the exons. For example, the chicken ovalbumin gene contains eight exons and seven introns distributed over 7700 base pairs (7.7 kilo-bases or 7.7 kb) of genomic DNA. The exons total only about 1.9 kb, which means that about 75% of the gene consists of introns.

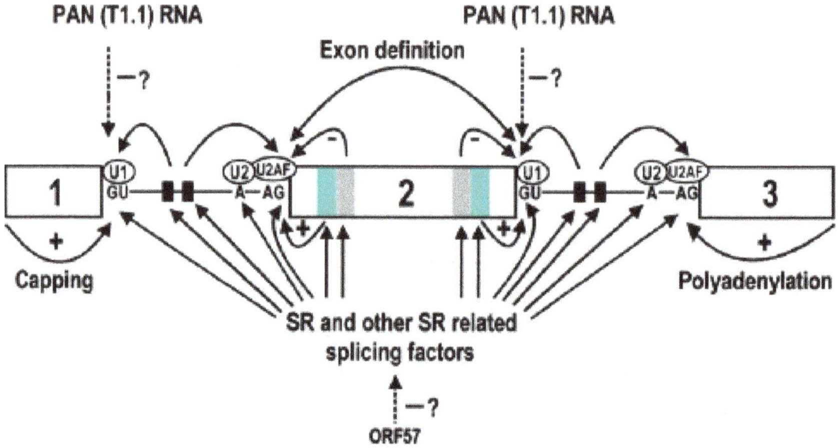

Fig. 2 Regulation of splice site selection by cis elements and trans splicing factors and possible involvement of KSHV gene products. Diagram shows a pre-mRNA containing three exons (open boxes) and two introns (lines). Two categories of splicing regulatory elements, splicing enhancers and splicing suppressors (silencers), have been described recently in the exons (grey boxes) or introns (small black boxes) of several viral and mammalian pre-mRNA substrates (94–99). The heavy- and light-grey boxes in exon 2 are, respectively, exonic splicing enhancers (ESE) and exonic splicing suppressor (ESS) which are often juxtaposed. The small black boxes in the introns are either intronic splicing enhancers (ISE) or intronic splicing suppressors (ISS). Through interacting with SR and other SR related splicing factors, these regulatory elements positively (+) or negatively (") control the selection of a splice site upstream or downstream. An internal exon is also usually defined as less than 500 nts in size (51). RNA 52 capping (100, 101) and 32 polyadenylation (102, 103) couple with RNA splicing to promote recognition of terminal splice sites. Negative effects of KSHV ORF57 and PAN RNA on RNA splicing are hypothetical.

A notable example is the human gene that encodes the blood clotting protein factor VIII. This gene contains about 186 kb of DNA and is divided into 26 exons. The mRNA is only about 9 kb long, which implies that about 175 kb of the gene contains introns. On average, introns constitute about 10 times more DNA than exons in the genes of higher eukaryotes.

Although most eukaryotic genes contain introns, their presence is not universal. Histone genes do not contain introns which shows that introns are not required for gene function in eukaryotic cells. Most genes of the simple eukaryote yeast lack introns. Introns are however, present in some rare genes of prokaryotes.

Most introns have no cellular function, although a few have been found to encode functional RNAs or proteins. In general, introns are considered to represent remnants of sequences that were important earlier in evolution. Introns may have helped to accelerate evolution by facilitating recombination between exons of different genes, the process is called exon shuffling.

Recombination between introns of different genes could produce new genes containing novel combinations of exons. This hypothesis is supported by studies that have demonstrated that some genes are chimaeras of exons derived from several other genes, thus indicating that new genes can be formed by recombination between intron sequences.

24

RECOMBINANT DNA, PCR AND DNA CHIPS / MICROARRAYS

RECOMBINANT DNA

Recombinant DNA technology involves using enzymes and various laboratory techniques to manipulate and isolate DNA segments of interest. This method can be used to combine (or splice) DNA from different species or to create genes with new functions. The resulting copies are often referred to as recombinant DNA. Such work typically involves propagating the recombinant DNA in a bacterial or yeast cell, whose cellular machinery copies the engineered DNA along with its own.

A technique mainly used to change the phenotype of an organism (host) when a genetically altered vector is introduced and integrated into the genome of the organism. So, basically, this process involves the introduction of a foreign piece of DNA structure into the genome which contains our gene of interest. This gene which is introduced is the recombinant gene and the technique is called the recombinant DNA technology.

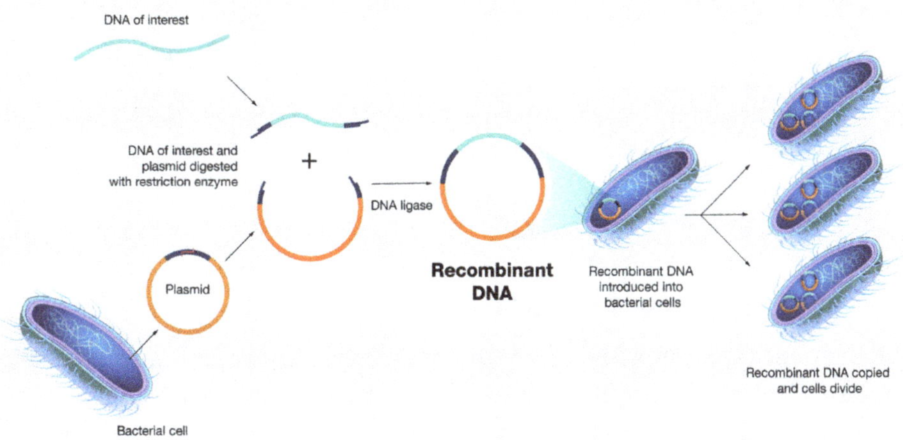

Fig. 1: Recombinant DNA

There are multiple steps, tools and other specific procedure followed in the recombinant DNA technology, which is used for producing artificial DNA to generate the desired product.

The technology used for producing artificial DNA through the combination of different genetic materials (DNA) from different sources is referred to as Recombinant DNA Technology. Recombinant DNA technology is popularly known as genetic engineering.

The recombinant DNA technology emerged with the discovery of restriction enzymes in the year 1968 by Swiss microbiologist Werner Arber,

Inserting the desired gene into the genome of the host is not as easy as it sounds. It involves the selection of the desired gene for administration into the host followed by a selection of the perfect vector with which the gene has to be integrated and recombinant DNA formed.

Thus the recombinant DNA has to be introduced into the host. And at last, it has to be maintained in the host and carried forward to the offspring.

Tools Of Recombinant DNA Technology

The enzymes which include the restriction enzymes help to cut, the polymerases- help to synthesize and the ligases- help to bind. The restriction enzymes used in recombinant DNA technology play a major role in determining the location at which the desired gene is inserted into the vector genome. They are two types, namely Endonucleases and Exonucleases.

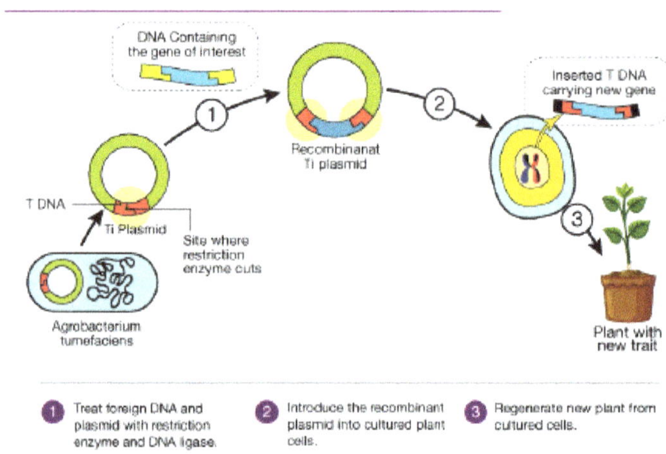

Fig. 2 Recombinant DNA Technology

The Endonucleases cut within the DNA strand whereas the Exonucleases remove the nucleotides from the ends of the strands. The restriction endonucleases are sequence-specific which are usually palindrome sequences and cut the DNA at specific points. They scrutinize the length of DNA and make the cut at the specific site called the restriction site. This gives rise to sticky ends in the sequence. The desired genes and the vectors are cut by the same restriction enzymes to obtain the complementary sticky notes, thus making the work of the ligases easy to bind the desired gene to the vector.

The vectors – help in carrying and integrating the desired gene. These form a very important part of the tools of recombinant DNA technology as they are the ultimate vehicles that carry forward the desired gene into the host organism. Plasmids and bacteriophages are the most common vectors in recombinant DNA technology that are used as they have a very high copy number. The vectors are made up of an origin of replication- This is a sequence of nucleotide from where the replication starts, a selectable marker – constitute genes which show resistance to certain antibiotics like ampicillin; and cloning sites – the sites recognized by the restriction enzymes where desired DNAs are inserted.

Host organism – into which the recombinant DNA is introduced. The host is the ultimate tool of recombinant DNA technology which takes in the vector engineered with the desired DNA with the help of the enzymes.

There are a number of ways in which these recombinant DNAs are inserted into the host, namely – microinjection, biolistics or gene gun, alternate cooling and heating, use of calcium ions, etc.

Process of Recombinant DNA Technology

The complete process of recombinant DNA technology includes multiple steps, maintained in a specific sequence to generate the desired product.

- **Step-1.** Isolation of Genetic Material.
- The first and the initial step in Recombinant DNA technology is to isolate the desired DNA in its pure form i.e. free from other macromolecules.
- **Step-2.**Cutting the gene at the recognition sites.
- The restriction enzymes play a major role in determining the location at which the desired gene is inserted into the vector genome. These reactions are called 'restriction enzyme digestions'.
- **Step-3.** Amplifying the gene copies through Polymerase chain reaction (PCR).
- It is a process to amplify a single copy of DNA into thousands to millions of copies once the proper gene of interest has been cut using the restriction enzymes.
- **Step-4.** Ligation of DNA Molecules.
- In this step of Ligation, joining of the two pieces – a cut fragment of DNA and the vector together with the help of the enzyme DNA ligase.
- **Step-5.** Insertion of Recombinant DNA Into Host.

In this step, the recombinant DNA is introduced into a recipient host cell. This process is termed as Transformation. Once after the insertion of the recombinant DNA into the host cell, it gets multiplied and is expressed in the form of the manufactured protein under optimal conditions.

As mentioned in Tools of recombinant DNA technology, there are various ways in which this can be achieved. The effectively transformed cells/organisms carry forward the recombinant gene to the offspring.

Application of Recombinant DNA Technology

- DNA technology is also used to detect the presence of HIV in a person.
- Gene Therapy – It is used as an attempt to correct the gene defects which give rise to heredity diseases.
- Clinical diagnosis – ELISA is an example where the application of recombinant
- Recombinant DNA technology is widely used in Agriculture to produce genetically-modified organisms such as Flavr Savr tomatoes, golden rice rich in proteins, Bt-cotton to protect the plant against ball worms and lot more.
- In the field of medicines, Recombinant DNA technology is used for the production of Insulin.

DNA Cloning

A clone is a cluster of individual entities or cells that are descended from one progenitor. Clones are genetically identical as the cell simply replicates producing identical daughter cells every time. Scientists are able to generate multiple copies of a single fragment of DNA, a gene which can be used to create identical copies constituting a DNA clone. DNA cloning takes place through the insertion of DNA fragments into a tiny DNA molecule. This molecule is made to replicate within the living cell, for instance, a bacterium. The tiny replicating molecule is known as the carrier of the DNA vector

Yeast cells, viruses, Plasmids are the most commonly used vectors. Plasmids are the circular DNA molecules that are introduced from bacteria. They are not part of the main cellular genome. It carries genes, which provide the host cell with beneficial properties such as mating ability, drug resistance. They can be conveniently manipulated as they are small enough and they are capable of carrying extra DNA which is weaved into them.

Applications of Gene Cloning

Listed below are the applications of gene cloning:

- Gene Cloning plays an important role in the medicinal field. It is used in the production of hormones, vitamins and antibiotics.
- Gene cloning finds its applications in the agricultural field. Nitrogen fixation is carried out by cyanobacteria wherein desired genes can be used to enhance the

productivity of crop and improvement of health. This practice reduces the use of fertilizers hence chemical-free produce is generated

- It can be applied to the science of identifying and detecting a clone containing a particular gene which can be manipulated by growing in a controlled environment
- It is used in gene therapy where a faulty gene is replaced by insertion of a healthy gene. Medical ailments such as leukaemia and sickle cell anaemia can be treated with this principle.

POLYMERASE CHAIN REACTION

Polymerase chain reaction or PCR is a reaction that is utilised to amplify a gene or fragment of DNA of interest. It is done in vitro using a primer. This technique is used in labs to make billions of copies of the desired gene for research, diagnostic and therapeutic purposes.

PCR was invented by Kary Mullis in 1983. PCR requires a DNA primer designed for the DNA template and DNA polymerase, preferably a thermostable DNA polymerase. In the polymerase chain reaction, the reaction cycle is repeated many times to get billions of copies.

PCR is an integral part of biotechnology, medical biology, diagnostics, forensic analysis, molecular biology research, etc. The amplified DNA can be sequenced, cloned and visualised by gel electrophoresis.

Polymerase Chain Reaction Steps

Each cycle of polymerase chain reaction has three steps. They are:

1. **Denaturation:** The first step in PCR is denaturation. Denaturation is required to separate the double-stranded DNA sample. It is done at 94-98°C for 20-30 seconds. It breaks the hydrogen bonds present between base pairs. Denaturation leads to the formation of single strands of DNA.
2. **Annealing:** The second step is the annealing of the primer. Here the reaction temperature is lowered to allow the complementary base pairing between the primer and the complementary part of the single strands of the DNA template. A proper temperature needs to be maintained in order to allow highly specific and proper primer hybridisation. Then DNA polymerase binds to the template-primer hybrid and starts the DNA synthesis.
3. **Extension:** A thermostable DNA polymerase is used for this purpose. Taq polymerase is commonly used for this purpose. It is done at a temperature of 75-80C (72°C). The DNA polymerase adds nucleotides in the 5'-3' direction and synthesises the complementary strand of the DNA template.

This cycle is repeated 25-30 times and with this, the DNA sample can be amplified a billion times. Gel electrophoresis is used to visualise the result of polymerase chain reaction.

PCR reaction requires a primer, thermostable DNA polymerase, template DNA and nucleotides.

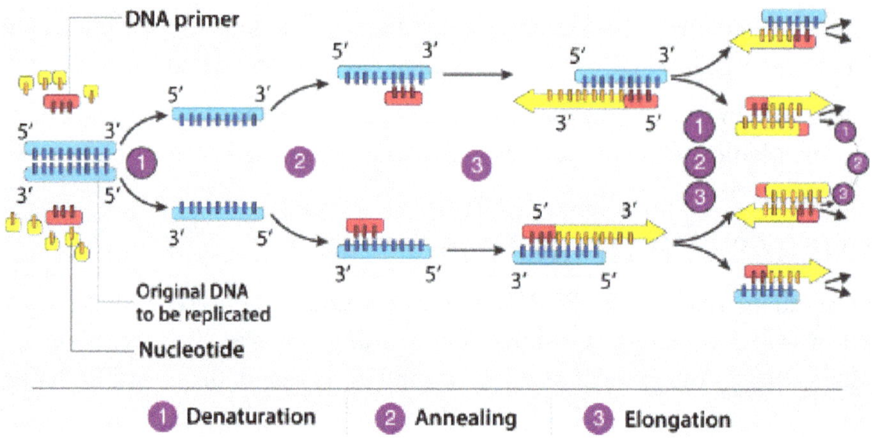

Fig. 24.3 Polymerase Chain Reaction

Primers for Polymerase Chain Reaction

Two sets of primers are used, which are short stretches of oligonucleotides. They are chemically synthesised and complementary to the part of the DNA template to be amplified. DNA polymerase requires a primer for initiating replication. DNA polymerase cannot initiate the process, it can only add nucleotides, therefore a primer is used for this purpose to initiate the poly-merisation process.

Primers are usually 20 nucleotides long. The two primers used, cover the target region of DNA to be amplified and bind to the opposite edges of the template strand. After binding of primers, the region between them gets copied by the DNA polymerase.

Taq Polymerase

The replication or polymerisation of DNA requires a DNA polymerase enzyme. In PCR, Taq polymerase is used. It is thermostable and is isolated from a heat-tolerant bacterium, *Thermus aquaticus*. A thermostable polymerase is required so that it can remain active at higher temperatures. Higher temperatures are used in the PCR process for denaturation of the DNA double helix repeatedly for multiple cycles.

Application of Polymerase Chain Reaction

Polymerase chain reaction has many advantages. It is a highly sensitive technique and gives results rapidly. It is simple and easy to understand. It is highly specific and is an important tool in research and biotechnology. Millions and billions of copies of the desired gene can be produced, which can be utilised for various purposes such as sequencing, analysis and cloning.

Some of the important applications of polymerase chain reaction are given below:

- PCR is used for diagnostics, e.g. A viral genome can be amplified and detected for the presence in the host cell.
- A genetic sequence associated with a bacterial toxin can be diagnosed efficiently using PCR, e.g. Pertussis toxin
- PCR amplification enables the detection of a viral genome early after infection. It can help in diagnosing the infection even before the onset of the disease. It significantly helps in the treatment of the disease.
- Viral load can be detected by PCR based techniques.
- RT-PCR is commonly used to detect the SARS-CoV-2 viral genome. It is a variant of PCR and uses the enzyme reverse transcriptase to produce DNA complementary to the viral RNA.
- Oncogenes mutations can be detected early with PCR. It helps in diagnosing various cancers such as leukaemia and lymphoma.
- PCR is used in forensics. A small sample of DNA collected from the crime scene can be amplified, analysed and compared with the suspects.
- PCR can be used for genetic testing. A gene associated with the genetic disorder can be amplified and detected from the sample taken from patients and also from foetal cells.
- PCR is used readily in molecular biology research. PCR helps in sequencing a genome. The unknown viral genome can be sequenced and studied for a better understanding of disease progression.
- PCR is utilised in recombinant DNA technology. The desired gene is amplified and then can be inserted into a vector for cloning.
- PCR is useful in detecting and analysing ancient DNA samples that are thousands of years old, e.g. forty-thousand-year-old mammoth, Egyptian mummies, etc.
- PCR is also helpful in analysing phylogenetic relationships from ancient samples.
- PCR is used to study gene expression. Alteration of gene expression can be studied in the case of tumours, cancers and other diseases.
- PCR can be used to create mutant genes of choice to study site-directed mutagenesis.
- PCR analysis is an important tool to monitor the spread of diseases. The new mutant forms present in the population can be detected and monitored.
- PCR is useful in prenatal testing for genetic disorders and also to find if a person is a carrier of a certain disease.

- Quantitative PCR or qPCR (Real-Time PCR) helps to determine the quantitative level of DNA or gene expression. Accumulation of DNA products is quantified after each cycle of PCR.

DNA MICROARRAY TECHNOLOGY

Scientists know that a mutation - or alteration - in a particular gene's DNA may contribute to a certain disease. However, it can be very difficult to develop a test to detect these mutations, because most large genes have many regions where mutations can occur. For example, researchers believe that mutations in the genes *BRCA1* and *BRCA2* cause as many as 60 percent of all cases of hereditary breast and ovarian cancers. But there is not one specific mutation responsible for all of these cases. Researchers have already discovered over 800 different mutations in *BRCA1* alone.The DNA microarray is a tool used to determine whether the DNA from a particular individual contains a mutation in genes like *BRCA1* and *BRCA2*. The chip consists of a small glass plate encased in plastic. Some companies manufacture microarrays using methods similar to those used to make computer microchips. On the surface, each chip contains thousands of short, synthetic, single-stranded DNA sequences, which together add up to the normal gene in question, and to variants (mutations) of that gene that have been found in the human population.

DNA microarray use

When they were first introduced, DNA microarrays were used only as a research tool. Scientists continue today to conduct large-scale population studies - for example, to determine how often individuals with a particular mutation actually develop breast cancer, or to identify the changes in gene sequences that are most often associated with particular diseases. This has become possible because, just as is the case for computer chips, very large numbers of 'features' can be put on microarray chips, representing a very large portion of the human genome.

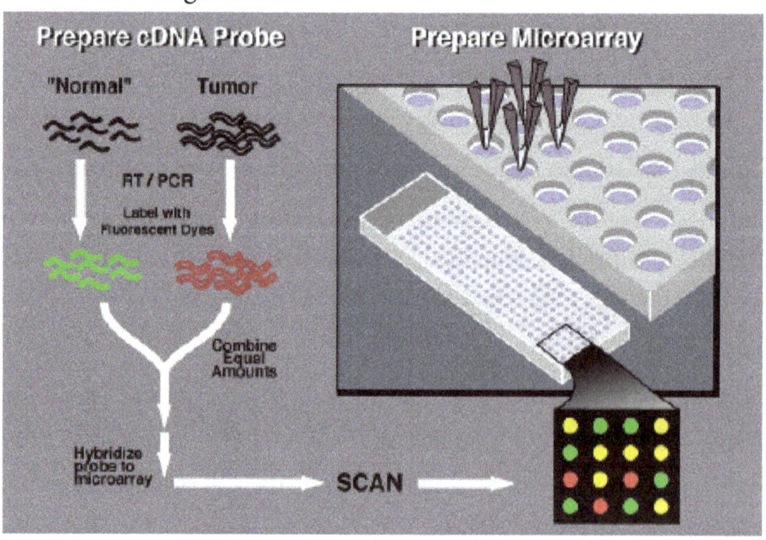

Fig. 4 Microarray Technology

Microarrays can also be used to study the extent to which certain genes are turned on or off in cells and tissues. In this case, instead of isolating DNA from the samples, RNA (which is a transcript of the DNA) is isolated and measured.

Today, DNA microarrays are used in clinical diagnostic tests for some diseases. Sometimes they are also used to determine which drugs might be best prescribed for particular individuals, because genes determine how our bodies handle the chemistry related to those drugs.With the advent of new DNA sequencing technologies, some of the tests for which microarrays were used in the past now use DNA sequencing instead. But microarray tests still tend to be less expensive than sequencing, so they may be used for very large studies, as well as for some clinical tests.

How does a DNA microarray work?

To determine whether an individual possesses a mutation for a particular disease, a scientist first obtains a sample of DNA from the patient's blood as well as a control sample - one that does not contain a mutation in the gene of interest.

The researcher then denatures the DNA in the samples - a process that separates the two complementary strands of DNA into single-stranded molecules. The next step is to cut the long strands of DNA into smaller, more manageable fragments and then to label each fragment by attaching a fluorescent dye (there are other ways to do this, but this is one common method). The individual's DNA is labeled with green dye and the control - or normal - DNA is labeled with red dye. Both sets of labeled DNA are then inserted into the chip and allowed to hybridize - or bind - to the synthetic DNA on the chip.

If the individual does not have a mutation for the gene, both the red and green samples will bind to the sequences on the chip that represent the sequence without the mutation (the "normal" sequence).

If the individual does possess a mutation, the individual's DNA will not bind properly to the DNA sequences on the chip that represent the "normal" sequence but instead will bind to the sequence on the chip that represents the mutated DNA.

25

SYNTHESIS, ISOLATION AND SEQUENCING OF GENES

SYNTHESIS OF GENES

Artificial gene synthesis is the chemical synthesis of a DNA sequence that represents one or more genes. While site-directed mutagenesis is regularly used to probe biological hypotheses by incorporating single base mutations, artificial gene synthesis provides a method to efficiently produce long stretches of natural and non-natural nucleic acid sequences, broadening the scope of biological experiments. Sequences that are hard to isolate from natural sources can be routinely generated in the lab, even entirely non-natural gene sequences can be synthesised. Not only does gene synthesis have implications for synthetic biology, it also introduces the possibility of developing genes containing modified nucleotides or even novel base pairs that could allow for the expansion of the genetic code.

Therefore, large-scale low-cost methods of gene synthesis must be readily available to facilitate relevant academic and commercial biological research.

Artificial gene synthesis

An initial challenge of gene synthesis was the production of long stretches of nucleotide sequence: genes can range from several hundred to several thousand base pairs in length. Current solid-phase oligonucleotide synthesis technology is limited to the generation of oligos 200 nucleotides in length. Two major methods exist to circumvent this issue: polymerase cycling assembly and synthesis by ligation.

Polymerase cycling assembly (PCA)

Polymerase cycling assembly (PCA) exploits single stranded templates rather than relying on total synthesis of both DNA strands. Short oligos are synthesised that alternate between both strands of the gene, overlapping by regions of 20-30 base pairs. Then, to give a superficial explanation, a polymerase uses dNTPs to "fill in the gaps" on one gene strand using the complementary bases on the other. A subsequent, entirely separate polymerase step then amplifies the constructed gene by traditional polymerase chain reaction (PCR).

With PCA, the commonly used explanation of the polymerase "filling in the gaps" is misleading: the gene is actually assembled step by step. Overlapping oligos anneal at their termini to form discrete overhanging dimers. The overhangs act as templates to elongate the complementary oligo by polymerase action generating fully double-stranded DNA that makes up a small section of the gene. These extended dsDNA sections are then melted and annealed to another overlapping oligo further along the gene, forming more single stranded overhangs. Further polymerase action turns these dimers into longer dsDNA fragments and this process continues until the entire gene is assembled.

For examples of any more than 4 oligos, the various iterations of annealing and elongation get too complicated to represent diagrammatically. One consistent point however is that polymerases extend 5' to 3' only, so although many oligo dimers will overlap and anneal, only those with the potential to extend 5' to 3' will be elongated. This limitation implies that the terminal oligos that cover the ends of the gene must be on the gene strand with the 5' terminus. Otherwise, if a terminal oligo were on the strand with the 3' terminus, the complementary oligo would have to be extended 3' to 5' to complete the gene, which is impractical for standard polymerase methodologies.

PCA eliminates the need for the whole gene to be synthesised solid phase. Once assembled, the gene product can be amplified by traditional PCR to generate a suitable concentration and purified by filtration to remove any remaining primers.

Error correction

Unfortunately, the process is not without errors. Insertion and deletion mutations (frameshift mutations) can be introduced at the solid phase synthesis stage and chain extension by polymerase risks further substitutions, insertions and deletions. Fortunately, natural DNA repair enzymes exist to correct these errors. However, since synthetic DNA has no indication of which strand has the correct sequence (unlike DNA in vivo), nucleases that rely on positive identification of the correct template strand cannot be used. Instead, it is necessary to use endonucleases that completely excise stretches of DNA that contain a mutation.

Since PCR cannot discriminate between correct sequences and those with errors, any mutations that have accumulated in the solid phase synthesis or polymerase extension of one gene strand will be matched in the reverse complement during amplification: the mutation is "invisible". Therefore, the mixture of correct and incorrect gene products is melted and reannealed. It is statistically unlikely that mutations will occur in the same place in the same way on two random strands. This mismatch perturbs the secondary structure of the gene.

Certain mismatch repair enzymes such as T7 Endonuclease I recognise these structural perturbations and cleave the first, second or third phosphodiester bond 3' to the mismatch (dependent on the specific enzyme) on both strands. A single strand exonuclease then chews up the overhanging single stranded DNA 3' to 5'. This process produces dsDNA of varying sizes that all overlap and between them contain the correct sequence spanning the entire gene. Subsequent PCA and PCR amplification generates the corrected gene product.

Gene synthesis by ligation

While PCA facilitates the synthesis of large genes and genomes, even after error correction the product often has many errors present in the sequence. Gene synthesis by ligation has a lower error rate than PCA, but is limited to the synthesis of smaller genes since the fragment oligos must span the entire sequence of both the sense and antisense strands: this is impractical for longer genes.

However, if the goal is to synthesise a smaller gene (< 2 kbp), gene synthesis by ligation will likely generate a product with fewer errors because:

- It removes the need for a polymerase that can often introduce mutations;
- Strands are subject to hybridisation selection - since all oligos overlap entirely, errors in their syntheses will produce mismatches which mean they are less likely to anneal and form a ligated product.

Solid phase synthesis generates the constituent fragments which are then purified ready for ligation. These oligos span the entire gene on both strands, missing only the phosphodiester bonds that link neighbouring oligos. Fragments are designed so two adjacent oligos are held together with a complementary template (the splint) then ligated enzymatically.

Three major methods of ligation-based synthesis exist :

- Shotgun ligation - the gene is synthesised en masse by simply mixing the constituent oligos together and adding a ligase. The reaction can occur at room temperature though it usually requires extended reaction times. Use of thermostable ligases and high purity oligos have now made this one-pot reaction more efficient.
- ligase chain reaction (LCR) - LCR originally superseded shotgun ligation as a more efficient method with shorter reaction times and the ability to ligate longer genes (still limited to 2 kbp). This method follows a similar set of melting and annealing steps to PCR though differs by using a ligase instead of a polymerase to ligate templated oligos until the entire gene is assembled.
- Solid-support based ligation-mediated oligonucleotide assembly - The initial dsDNA oligo is immobilised on a solid support and iterative rounds of annealing and ligation steadily assemble the entire gene, which is cleaved from the support at the end of the synthesis.

While the error rate of ligation based synthesis is lower, it may still be necessary to perform error correction to yield a legitimate number of viable genes. Furthermore, PCA is the dominant methodology used for artificial gene synthesis. Ligation-based techniques require polyacrylamide gel electrophoresis (PAGE) purification and 5' phosphorylation, making the whole method costly and labour intensive. PCA and associated error correction is preferred as these steps are eliminated.

Further methods of gene synthesis exist including assembly directly on to the plasmid relying on bacterial homologous recombination (Sequence- and Ligation- Independent Cloning - SLIC) or in vivo assembly of transformed oligonucleotides by the yeast Saccharomyces cerevisiae. PCA and ligation based methods remain by far the most common methods of gene synthesis.

ISOLATING GENES

Earlier in this chapter, we discussed methods such as column chromatography that are used to purify proteins of interest. Using combinations of these methods, it is possible to isolate a protein to a high degree of purity, thus enabling us to study the protein's activity and properties. This problem is harder to solve for nucleic acids. Genomic DNA can be readily obtained from cells, but is too complex to be analyzed as a whole. Individual genes are the units of DNA that correspond to proteins, and thus, it makes more sense to isolate specific genes for study. Methods to isolate genes were not available till the 1970s, when the discovery of restriction enzymes and the invention of molecular cloning provided, for the first time, ways to obtain large quantities of specific DNA fragments, for study. Although, for purposes of obtaining large amounts of a specific DNA fragment, molecular cloning has been largely replaced by direct amplification using the polymerase chain reaction described later, cloned DNAs are still very useful for a variety of reasons. The development of molecular cloning was dependent on the discovery of restriction endonucleases, described below.

Restriction enzymes

Restriction enzymes, or restriction endonucleases, are enzymes made by bacteria. These enzymes protect bacteria by degrading foreign DNA molecules that are carried into their cells by, for example, an invading bacteriophage. Each restriction enzyme recognizes a specific sequence, usually of four or six nucleotides in the DNA. These sequences, when they occur in the bacterium's own DNA, are chemically modified by methylation, so that they are not recognized and degraded. Where these sequences occur in foreign DNA, they are cut by the restriction enzyme.

The utility and importance of restriction enzymes lies in their ability to recognize specific sequences in DNA and cut near or (usually) at the site they recognize. Over 3000 such enzymes are known. Sequences recognized by these enzymes are typically 4-8 base pairs long and the most commonly used enzymes recognize sequences described as palindromic.

Palindrome

In molecular biology, the term palindrome means that the sequence of the recognition site when read in the 5‘ to 3‘ direction for the top strand is exactly the same as that of the bottom strand. Consider the sequence recognized by the restriction enzyme known as Hind III (pronounced hin-dee-three). It is

5’ -A-A-G-C-T-T-3’

3’ -T-T-C-G-A-A-5’

On the top strand, the recognition sequence is

5’ AAGCTT 3’

which is the same as the bottom strand (read in the same 5’ to 3’ direction).

While all restriction enzymes must recognize and bind to particular DNA sequences, the exact spot at which they cut the DNA varies. Some enzymes leave a staggered sequence after cutting that has an overhang at the 5’ end of one strand of the duplex; some leave a staggered sequence after cutting that has an overhang at the 3’ end; and some cut both strands in the same place, leaving no overhanging sequence - called blunt end cutters.

Consider cutting a DNA sequence that contains the Hind III recognition site, which is

5’ -A-A-G-C-T-T-3’

3’ -T-T-C-G-A-A-5’

Embedded within a DNA sequence, the Hind III sequence would look like this (Ns correspond to any base and represent all of the DNA around the recognition site).

5’ -N-N-N-A-A-G-C-T-T-N-N-N-3’

3’ -N-N-N-T-T-C-G-A-A-N-N-N-5’

After cutting with Hind III, it would look as follows:

5’ -N-N-N-A 3‘ 5’A-G-C-T-T-N-N-N-N-3’

3’ -N-N-N-T-T-C-G-A-5‘ 3’ A-N-N-N-N-5’

Where gaps have been inserted to illustrate where cutting has occurred. Hind III cuts etween the two ‘A’ containing nucleotides near the 5’ end of the recognition sequence and thus leaves 5’ overhangs (Figure 1).

The restriction enzyme Pst I, on the other hand, recognizes the following sequence

5’ -N-N-N-C-T-G-C-A-G-N-N-N-N-3’

3’ -N-N-N-G-A-C-G-T-C-N-N-N-N-5’

And cuts between the A and the G near the 3’ end of the recognition sequence.

5’ -N-N-N-C-T-G-C-A 3‘ 5’G-N-N-N-N 3’

3’ -N-N-N-G 5‘ 3’ A-C-G-T-C-N-N-N-N 5’

As you can see, cutting a DNA with Pst I leaves 3' overhangs of the recognition sequence. The ends left after cutting by a restriction enzyme that overhang either at the 5' end or the 3' end are referred to as being "sticky" because they can form proper base pairs and more readily be joined to a similarly "sticky end". This means that you can take two unrelated pieces of DNA, cut them with the same restriction enzyme so that they have compatible sticky ends, and then "paste" them together using DNA ligase to form a new hybrid molecule, or recombinant.

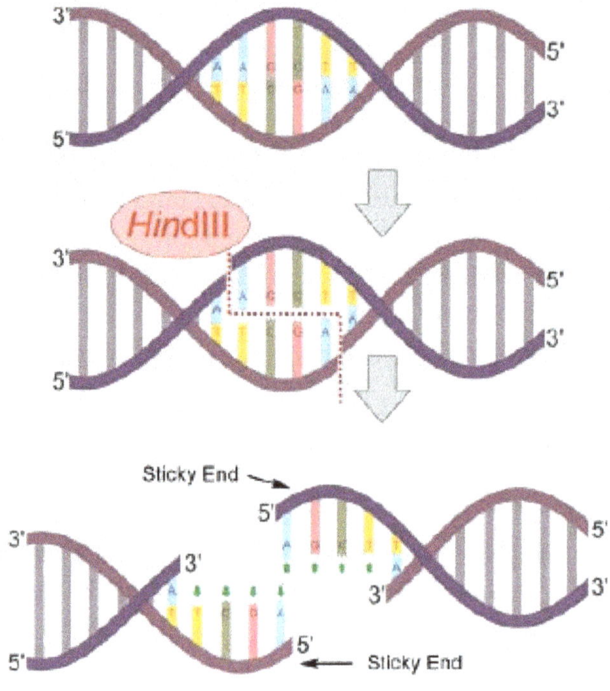

Fig. 1 Result of cutting DNA with Hind III

Making Recombinant DNAs

Joining together of DNA fragments from different sources creates recombinant DNA. The ability to cut and paste DNA might seem like purely a technical feat, but one key application that arose out of this is molecular cloning. In molecular cloning a gene of interest can be inserted into a vector, usually a plasmid, by cutting both the vector and the gene (called the insert) with the same enzyme to generate sticky ends and joining the two pieces together to generate a recombinant (Fig. 2). A plasmid is a type of autonomously replicating, extrachromosomal DNA. It is quite simple to extract plasmids from the cells, engineer them to contain the gene of interest and re-introduce the recombinant plasmid into the bacteria. The idea was that when the plasmid DNA was replicated, the extra inserted gene would also be copied. Thus, by growing up a lot of the bacteria carrying the plasmid, many copies of the gene of interest could be obtained, to provide sufficient amounts of the gene to use in experiments. While we now have easier methods to accomplish this goal, cloned DNAs remain very useful. For example, it is possible to clone a gene that encodes a protein of interest so that it can be expressed at high levels in the cells into which the recombinant plasmid is introduced.

Whatever the purpose for which the recombinant plasmid is made, it typically carries an antibiotic resistance gene (or genes), called a selectable marker. Cells that take up the plasmid will be able to grow in the presence of the antibiotic. If bacterial cells to which the plasmid has been added are plated on agar containing the antibiotic, the cells which took up the plasmid will be able to grow, while the others will not.

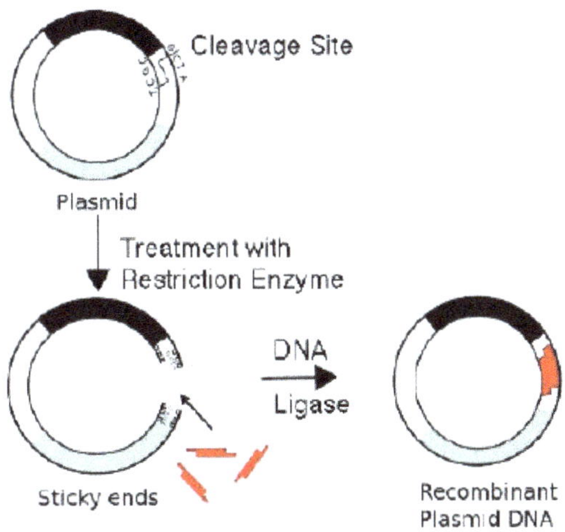

Fig. 2: Recombinant DNA construction

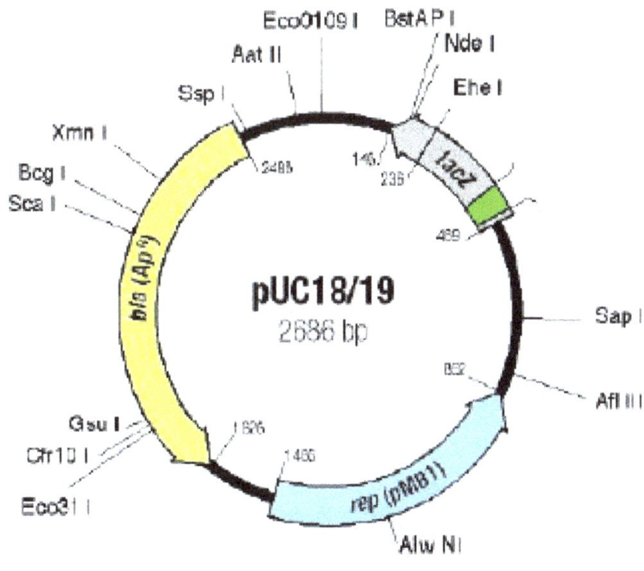

Fig. 3: Restriction site map for the pUC 18/19 plasmids, a classic plasmid vector. Genes identified by arrows. Numbers correspond to the pUC 18/19 numbering convention

SEQUENCING OF GENES

DNA Sequencing Techniques

DNA sequencing is the process of determining the nucleic acid sequence – the order of nucleotides in DNA. It includes any method or technology that is used to determine the order of the four bases: adenine, guanine, cytosine, and thymine. The advent of rapid DNA sequencing methods has greatly accelerated biological and medical research and discovery.

Knowledge of DNA sequences has become indispensable for basic biological research, and in numerous applied fields such as medical diagnosis, biotechnology, forensic biology, virology and biological systematics. Comparing healthy and mutated DNA sequences can diagnose different diseases including various cancers, characterize antibody repertoire, and can be used to guide patient treatment. Having a quick way to sequence DNA allows for faster and more individualized medical care to be administered, and for more organisms to be identified and cataloged.

The rapid speed of sequencing attained with modern DNA sequencing technology has been instrumental in the sequencing of complete DNA sequences, or genomes, of numerous types and species of life, including the human genome and other complete DNA sequences of many animal, plant, and microbial species.

The first DNA sequences were obtained in the early 1970s by academic researchers using laborious methods based on two-dimensional chromatography. Following the development of fluorescence-based sequencing methods with a DNA sequencer, DNA sequencing has become easier and orders of magnitude faster.

The canonical structure of DNA has four bases: thymine (T), adenine (A), cytosine (C), and guanine (G). DNA sequencing is the determination of the physical order of these bases in a molecule of DNA. However, DNA bases are often modified by epigenetic processes to control gene expression. Thus, many other modified bases that may be present in a DNA molecule than the standard four bases. For example, in some viruses (specifically, bacteriophage), cytosine may be replaced by hydroxymethyl- or hydroxymethylglucose cytosine. In eukaryotic DNA, variant bases with methyl groups or phosphosulfate may be found as shown in Fig. 4. Depending on the sequencing technique, a particular modification, e.g., the 5mC (5'-methylcytosine) common in humans, may or may not be detected.

Early DNA sequencing methods

The first method for determining DNA sequences involved a location-specific primer extension strategy established by Ray Wu at Cornell University in 1970. DNA polymerase catalysis and specific nucleotide labeling, both of which figure prominently in current sequencing schemes, were used to sequence the cohesive ends of lambda phage DNA.

Between 1970 and 1973, Wu, R Padmanabhan and colleagues demonstrated that this method can be employed to determine any DNA sequence using synthetic location-specific primers. Frederick Sanger then adopted this primer-extension strategy to develop more rapid DNA sequencing methods at the MRC Centre, Cambridge, UK and published a method for "DNA sequencing with chain-terminating inhibitors" in 1977. Walter Gilbert and Allan Maxam at Harvard also developed sequencing methods, including one for "DNA sequencing by chemical degradation". In 1973, Gilbert and Maxam reported the sequence of 24 basepairs using a method known as wandering-spot analysis. Advancements in sequencing were aided by the concurrent development of recombinant DNA technology, allowing DNA samples to be isolated from sources other than viruses.

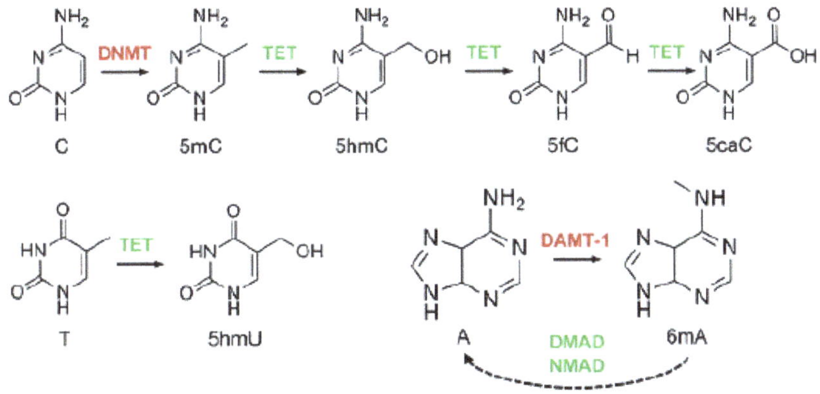

Fig. 4: DNA modifications with epigenetic regulatory functions and their interdependencies.

Maxam-Gilbert sequencing requires radioactive labeling at one 5' end of the DNA and purification of the DNA fragment to be sequenced. Chemical treatment then generates breaks at a small proportion of one or two of the four nucleotide bases in each of four reactions (G, A+G, C, C+T). The concentration of the modifying chemicals is controlled to introduce on average one modification per DNA molecule. Thus a series of labeled fragments is generated, from the radiolabeled end to the first "cut" site in each molecule. The fragments in the four reactions are electrophoresed side by side in denaturing acrylamide gels for size separation. To visualize the fragments, the gel is exposed to X-ray film for autoradiography, yielding a series of dark bands each corresponding to a radiolabeled DNA fragment, from which the sequence may be inferred.

The technical aspects of Maxam-Gilbert sequencing caused it to go out of favor once the Sanger sequencing method had been well established, as described below.

Sanger Sequencing Method

The chain-termination method developed by Frederick Sanger and coworkers in 1977 soon became the method of choice, owing to its relative ease and reliability. When invented, the chain-terminator method used fewer toxic chemicals and lower amounts of radioactivity than the Maxam-Gilbert method. Because of its comparative ease, the Sanger method was soon automated and was the method used in the first generation of DNA sequencers.

The classical chain-termination method requires a single-stranded DNA template, a DNA primer, a DNA polymerase, normal deoxynucleotide-triphosphates (dNTPs), and modified di-deoxynucleotidetriphosphates (ddNTPs), the latter of which terminate DNA strand elongation. These chain-terminating nucleotides lack a 3'-OH group required for the formation of a phosphodiester bond between two nucleotides, causing DNA polymerase to cease extension of DNA when a modified ddNTP is incorporated. The ddNTPs may be radioactively or fluorescently labelled for detection in automated sequencing machines.

The DNA sample is divided into four separate sequencing reactions, containing all four of the standard deoxynucleotides (dATP, dGTP, dCTP and dTTP) and the DNA polymerase. To each reaction is added only one of the four dideoxynucleotides (ddATP, ddGTP, ddCTP, or ddTTP).

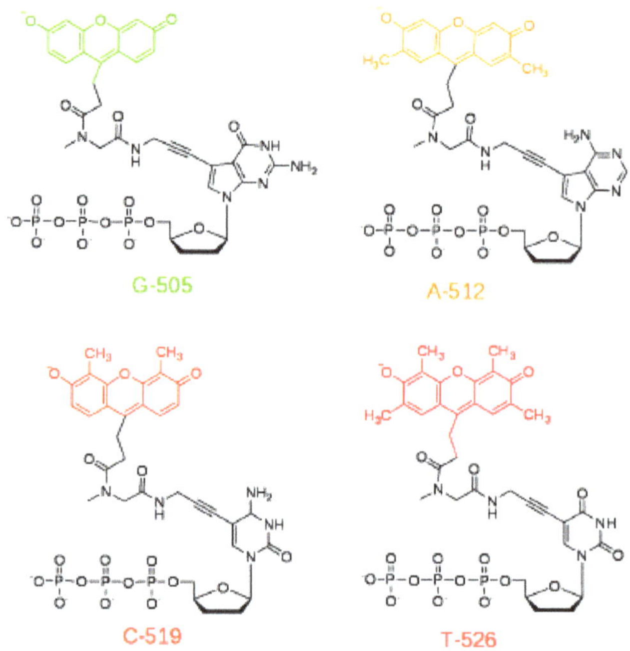

Fig. 5: Fluorescent ddNTPs for Sanger Sequencing. Dideoxynucleotides are utilized for sequencing as they cannot be extended further once they are incorporated into the nacent DNA.

The dideoxynucleotide concentration should be approximately 100-fold lower than that of the corresponding deoxynucleotide (e.g. 0.005mM ddTTP : 0.5mM dTTP) to allow enough fragments to be produced while still transcribing the complete sequence.

In total, four separate reactions are needed in this process to test all four ddNTPs. The is illustrated in Fig. 6 below.

Following rounds of template DNA extension from the bound primer, the resulting DNA fragments are heat denatured and separated by size using gel electrophoresis. This technique was frequently performed using a denaturing polyacrylamide-urea gel with each of the four reactions run in one of four individual lanes (lanes A, T, G, C). The DNA bands may then be visualized by autoradiography or UV light and the DNA sequence can be directly read off the X-ray film or gel image, as shown in Fig. 7 below.

Automation of the Sanger sequencing method was made possible when the shift from radioactively tagged nucleotides to fluorescently tagged nucleotides was made. Within the automated sequencers, capillary gel electrophoresis is performed rather than separating the samples using gel electrophoresis. The output from capillary electrophoresis are fluorescent peak trace chromatograms.

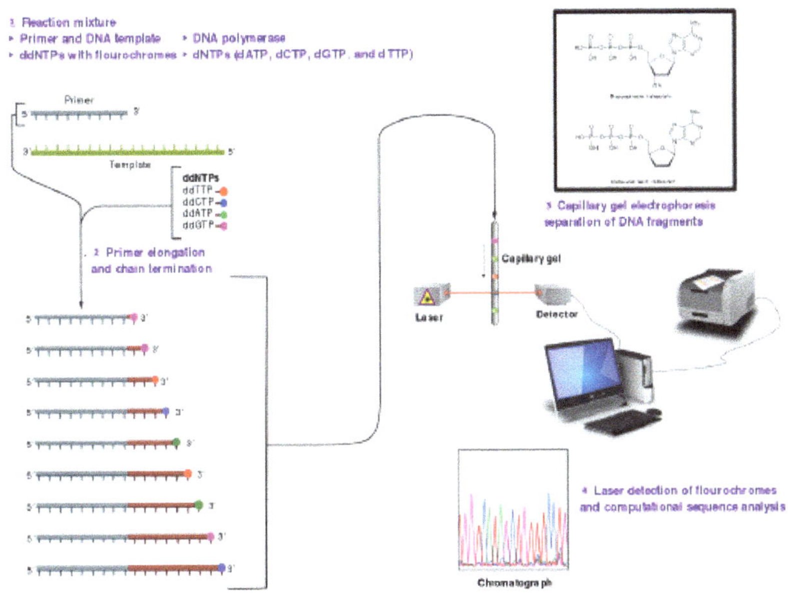

Fig. 6: The Sanger (chain-termination) method for DNA sequencing.

(1) A primer is annealed to a sequence, (2) Reagents are added to the primer and template, including: DNA polymerase, dNTPs, and a small amount of all four dideoxynucleotides (ddNTPs) labeled with fluorophores. During primer elongation, the random insertion of a ddNTP instead of a dNTP terminates synthesis of the chain because DNA polymerase cannot react with the missing hydroxyl. This produces all possible lengths of chains. (3) The products are separated on a single lane capillary gel, where the resulting bands are read by a imaging system. (4) This produces several hundred thousand nucleotides a day, data which require storage and subsequent computational analysis.

Automated DNA-sequencing instruments (DNA sequencers) can sequence up to 384 DNA samples in a single batch. Batch runs may occur up to 24 times a day greatly enhancing the speed with which samples may be sequenced and analyzed. Common challenges of DNA sequencing with the Sanger method include poor quality in the first 15-40 bases of the sequence due to primer binding and deteriorating quality of sequencing traces after 400-500 bases.

Sanger sequencing is the method which prevailed from the 1980s until ~2005. Over that period, great advances were made in the technique, such as fluorescent labeling, capillary electrophoresis, and general automation. These developments allowed much more efficient sequencing, leading to lower costs. The Sanger method, in mass production form, is the technology which produced the first human genome in 2001, ushering in the age of genomics.

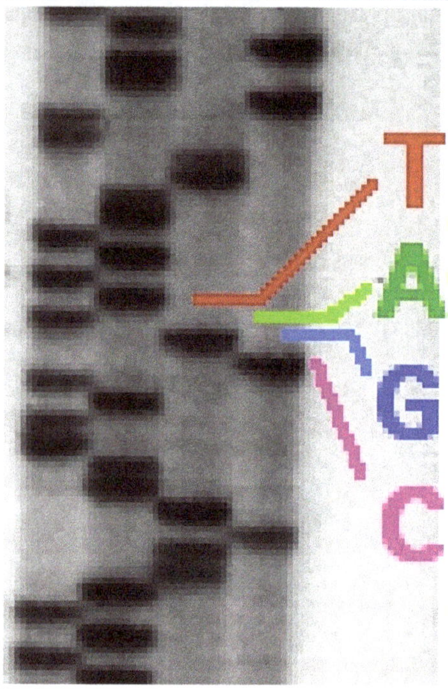

Fig. 7: Traditional Sanger Sequencing Gel.

Sequence visualized by autoradiography. Each lane contains a single reactions that has all four regular nucleotides and a small amount of one of the dideoxynucleotides (ddNTPs). Overtime, the ddNTPs will be incorporated at each position containing that specific nucleotide. The gel can then be read from the bottom to the top, as the smallest fragments.

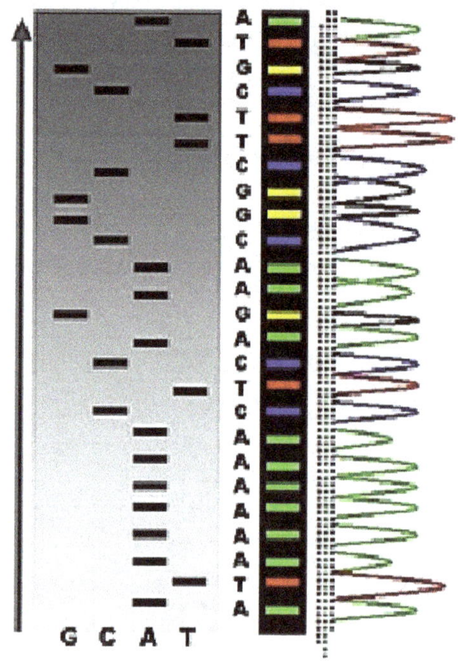

Fig. 8: Side by Side Comparison of Gel Electrophoresis and Capillary Electrophoresis.

Lefthand Diagram shows the traditional autoradiogram of Sanger sequencing samples. The Righthand Diagram shows the same reactions using fluorescently tagged ddNTPs separated by capillary electrophoresis. The chromatogram output is shown on the far right.

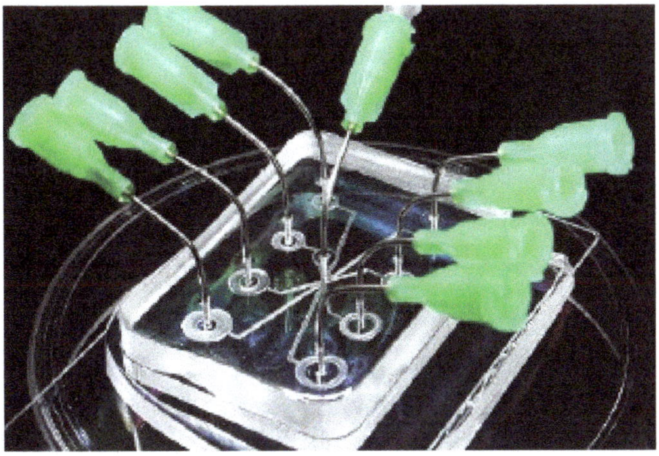

Fig. 9: Lab-On-A-Chip Technologies.

Microfluidic Sanger Sequencing

Microfluidic Sanger sequencing is a *lab-on-a-chip application* for DNA sequencing, in which the Sanger sequencing steps (thermal cycling, sample purification, and capillary electrophoresis) are integrated on a wafer-scale chip using nanoliter-scale sample volumes (Fig. 9). This technology generates long and accurate sequence reads, while obviating many of the significant shortcomings of the conventional Sanger method (e.g. high consumption of expensive reagents, reliance on expensive equipment, personnel-intensive manipulations, etc.) by integrating and automating the Sanger sequencing steps.

GLOSSARY

Active site The part of protein that must be maintained in a specific shape if the protein is to be functional.

Allele One of two copies of a gene.

Amino acid A peptide; the basic building block of proteins.

Amplicon Replicated target molecules created by polymerase chain reaction or other nucleic acid amplification methods.

Amplification Increasing the number of copies of a desired DNA segment. The amplified regions are called amplicons.

Aneuploid cell A cell having a chromosome number that differs from the normal chromosome number for the species by a small number of chromosomes.

Annealing Spontaneous alignment of two single DNA strands to form a double helix.

Antibody A protein molecule, produced by the immune system, that recognizes a particular substance and binds to it.

Anticipation (genetic) Apparent earlier age of onset and increased severity of a disease in successive generations, eg, Huntington's disease.

Anticodon A nucleotide triplet in a tRNA molecule that aligns with a particular codon in mRNA under the influence of the ribosome so that the amino acid carried by the tRNA is inserted in a growing protein chain.

Antiparallel A term used to describe the opposite orientations of the two strands of a DNA double helix; the 5' end of one strand aligns with the 3' end of the other strand.

Autoradiography The exposure of roentgenographic film to a blot or membrane containing a radiolabeled probe, used to locate the labeled probe

Avidin A protein that specifically binds to biotin with usually high affinity.

Bacteriophage A virus that infects bacteria

Base analog A chemical whose molecular structure mimics that of a DNA base; because of the mimicry, the analog may act as a mutagen.

Base pair (bp) One pair of complementary nucleotides (e.g. adenine to thymine or guanine to cytosine) on opposite sides of the duplex strands

Base One nucleotide consisting of a nucleoside, pentose sugar and triphosphate; building blocks for either DNA or RNA

cDNA Single-stranded complementary DNA, ie, a DNA molecule synthesized from a RNA template by reverse transcription of RNA.

Cell cycle The set of events that take place in the divisions of mitotic cells. The cell cycle oscillates between mitosis (M phase) and interphase. Interphase can be subdivided in order into G1, S phase and G2. DNA synthesis takes place during S phase. The length of the cell cycle is regulated through a special option in G1, in which G1 cells can enter a resting phase call G0

Chromosome Structure of DNA and associated proteins that contain the hereditary material within the cell. Genes are organized in linear arrangement in the chromosome.

Cistron Segments of DNA corresponding to one polypeptide chain, plus transcriptional start and stops signs.

Cloning The process of generating a large number of identical DNA fragments, typically to produce a probe for a specific gene.

Codominance The situation in which a heterozygote shows the phenotypic effects of both alleles.

Codon A sequence of three nucleotides that specifies a particular amino acid

Common disorder common variant (CDCV) hypothesis A theory that many common diseases are caused by common alleles that individually have little effect, but in concert confer a high risk.

Comparative genome hybridization (CGH) CGH is a molecular-cytogenetic method for the analysis of copy number changes (gains or losses) in the DNA content of a given individual's DNA.

Competent Able to take up exogenous DNA and thereby be transformed.

Complementarity The specific binding of adenine to thymidine (or uracil in RNA) and cytosine to guanine on opposite strands of DNA or RNA.

Complementary DNA or Copy DNA(cDNA) DNA generated form mRNA by the use of reverse transcriptase.

Complementary RNA Synthetic RNA produced by transcription from a specific DNA single-stranded template.

Complementation The production of a wild-type phenotype when two different mutations are combined in a diploid or a heterokaryon.

Complex disease A disorder in which the cause is considered to be a combination of genetic effects and environmental influences.

Compound heterozygosity Heterozygosity for two different mutant alleles of a gene, often the case for autosomal recessive disorders.

Copy number variation (CNV) A segment of DNA in which copy number differences have been found by comparison of two or more genomes. The segment may range from one kilobase to several megabases in size. The variation is usually due to deletion or duplication.

Cosmid A genetically engineered construct containing sites from lambda phage that allow for insertion of large pieces of DNA (50-30 kb). Recombinant cosmids can then be replicated in bacterial hosts.

Crossing over: Exchange of genetic material between two non sister chromatids takes place during meiosis cell division.

Homologous chromosome: A pair of chromosome having same size and shape bearing corresponding gene.

Phenotype: The character of organism which can be seen directly.

Genotype: Genetic constitution of organism.

Tonoplast: The membrane surrounding the vacuole.

Unit membrane: The basic trilamilar structure of cell membrane.

Deep resequencing A technique for sequencing a gene in several thousand subjects, typically using one of the new high-throughput sequencing technologies.

Degenerate code A genetic code in which some amino acids may be encoded by more than one codon each.

Degradation Due to physical shearing or exposure to endogenous or exogenously added nucleases, DNA and RNA can become hydrolyzed or degraded to the oligonucleotide or single-nucleotide level.

Denaturation The process of making double-stranded DNA single stranded.

Dideoxy sequencing A method of DNA sequencing in which dideoxynucleotide triphosphates (ddNTPs) are used in the growing oligonucleotide chains synthesized from the DNA template, thereby terminating elongation.

Disorder caused by one or more mutations in a single gene, eg, cystic fibrosis (mutations in the *CFTR* gene). Such disorders are also sometimes referred to Mendelian diseases.

DNA Deoxyribonucleic acid

DNA ligase An enzyme that covalently joins two pieces of double stranded DNA.

DNA polymerase The enzymes(s) that make DNA by the addition of bases to the end of the replicating DNA strand.Also has an editing function that can repair nicked DNA by removal of old nucleotides and replacement with new (e.g. labelled) nucleotides.

DNAase Deoxyribonuclease, an endonuclease that randomly hydrolyzes (nicks) DNA at single sites on either strand of the DNA

Dominant allele An allele that expresses its phenotypic effect even when heterozygous with a recessive allele; thus if A is a dominant over a, then A/A and A/a have the same phenotype.

Dominant phenotype The phenotype of a genotype containing the dominant allele; the parental phenotype that is expressed in a heterozygote.

Dot (slot) blot The immobilization of DNA or RNA to a sample spot (slot) on a filter that will be subsequently probed by hybridization. Usually many samples are applied, each to a different location on the same filter, for mass screening.

ds Double stranded

End-labeling The incorporation of nucleotides at the 5‘ end of a strand of DNA by the use of specific enzymes as one means of making labeled probes.

Endonuclease Enzyme that hydrolyzes dsDNA at internal locations

Enhancer A regulatory sequence that can elevate levels of transcription from an adjacent promoter. Many tissue-specific enhancers can determine patterns of gene expression in higher eukaryotes. Enhancers can act on promoters over many tens of kilobases of DNA and be 5’ or 3’ to the promoter that they regulate. Generally they are on the same chromosome.

Epigenetics Heritable changes to DNA structure that do not alter the underlying DNA sequence, eg, DNA methylation.

Epigenomics The application of epigenetics to the whole genome.

Ethidium bromide A chemical dye that intercalates between the bases in DNA and causes DNA to fluoresce when illuminated with ultraviolet light.

Euploid A cell having any number of complete chromosome sets or an individual composed of such cells.

Exome The approximately 1% of the human genome that comprises all exons and therefore the entire protein-coding region of the genome.

Exon The portion of a gene that is actually translated into protein. [Eukaryotic genes only.]

Exonuclease Enzyme that hydrolyzes ss or ds NA from the ends

Extracellular fluid (ECF): fluid exterior to cells; includes the interstitial fluid, blood plasma, and fluids found in other reservoirs in the body

Fingerprint The characteristic spot pattern produced by electrophoresis of the polypeptide fragments obtained through denaturation of a particular protein by a proteolytic enzyme.

FISH Fluorescent in situ hybridization. In situ (in the physical location) hybridization using a probe coupled to a fluorescent molecule.

Fluid compartment: fluid inside all cells of the body constitutes a compartment system that is largely segregated from other systems

Frame-shift mutation The insertion or deletion of a nucleotide pair or pairs, causing a disruption of the translational reading frame.

Functional genomics The study of patterns of gene expression and interaction in the genome as a whole.

Gene dose The number of copies of a particular gene present in the genome.

Gene locus The specific place on a chromosome where a gene is located.

Gene therapy The correction of a genetic deficiency in a cell by the addition of new DNA with or without its insertion into the genome.

Gene A sequence of nucleotides that code for a protein product.

Genetic association The nonrandom occurrence of a genetic marker (usually a particular allele of a polymorphism) with a trait, which suggests an association between the genetic marker (or marker close to it) and disease pathogenesis.

Genetic code The set of correspondences between nucleotide pair triplets in DNA and amino acids in proteins.

Genetics The study of genes.

Genome In eukaryotes, the basic (monoploid) chromosome set, consisting of a species-specific number of linkage groups and the genes contained therein. For example, in humans, the genome consists of the 24 different chromosomes (22 autosomes, X and Y chromosomes). The mitochondrial DNA is usually considered to be a separate "mitochondrial" genome.

Genome The complete set of hereditary factors of an organism, contained in the chromosome.

Genome-wide association study (GWAS) A test for the association between genetic polymorphisms spread evenly over the entire genome, and a disease. Usually at least 300 000 markers are required to adequately cover the genome.

Genomic library Collection of DNA fragments (library) whose sum represents the entire genomic DNA of an organism.

Genotype The genetic constitution with respect to the alleles at one or more pairs of genetic loci under observation. The genotype of an individual is the sum total of the genetic information contained on the chromosomes, as distinguished from the individual's phenotype (idiotype).

Genotype The specific genes that are present in an individual; they may or may not be expressed.

Germ line The cell lineage in a multi-tissued eukaryote from which the gametes derive.

Growth factor Signaling molecules, usually secreted polypeptides, that induce cell division in cells receiving theses signals. Signals can be autocrine (self derived), paracrine (from another cell but local in origin), endocrine (from a distant cell/ organ) or neurocrine (from nerve cells).

Haploid A single genome or set of chromosomes (eg, in human gametes, n=23), compared to the normal diploid (double) set of chromosomes (n=46).

Haploid A cell having one chromosome set or an organism composed of such cells.

Haploid: Single set of chromosome in cell is called haploid which is found in gametes.

Haplotype A combination of alleles at closely linked gene loci that are inherited together.

Hemizygous When one or more genes is present in only one, instead of two copies, eg, men are hemizygous for most genes on the X and Y chromosomes.

Heteroduplex A DNA double helix formed by annealing single strands of a heteroduplex DNA.

Heterozygote An individual having a heterozygous gene pair.

Heterozygous Having different alleles for one or more genes in homologous chromosome segments, as opposed to being homozygous with identical alleles at these loci.

Homology Similarity between two distinct genes in their nucleotide sequence.

Hybridization The process of complementary base pairing between two single strands of DNA, DNA and RNA, or sense and anti-sense RNA.

Hydrostatic pressure Pressure exerted by a fluid against a wall, caused by its own weight or pumping force

***In situ* hybridization** Use of labeled DNA or RNA probes to localize complementary sequences within a cell.

Intron A portion of a gene not translated into protein, even though it is transcribed into RNA. A splicing event removes it from the primary RNA transcript leaving only the exons in the mRNA. [Eukaryotes only.]

Inhibitor As it pertains to DNA manipulation in vitro, contamination in the preparation or sample can inhibit the biochemical processes involved in the manipulation, generally polymerases. Can be endogenous (from the sample matrix) or exogenous.

iInterstitial fluid (IF): fluid in the small spaces between cells not contained within blood vessels

Intracellular fluid (ICF): fluid in the cytosol of cells

Inverted repeat sequence A sequence found in identical form, but inverted. Commonly found in transposons and viruses.

Karyokinesis: Division of nucleus during cell division.

Cytokinesis: Division of cytoplasm.

Diploid: Two complete set of chromosome is known as diploid which is found in somatic cell.

Kilobase (kb) Unit of 1000 base pairs of DNA or 1000 bases of RNA.

Labeling The process of marking a DNA molecule using radioactive or non-radioactive labels.

Lagging strand In DNA replication, the strand that is synthesized apparently in the 3' to 5' direction, by ligating short fragments synthesized individually in the 5' to 3' direction. The short fragments called Okazaki fragments are primed using snRNAs.

Ligase chain reaction (LCR) An in vitro nucleic acid amplification method that uses DNA ligase, an enzyme that joins two pieces of DNA together.

Ligase An enzyme that can rejoin a broken phosphodiester bond in a nucleic acid.

Linkage disequilibrium (LD) Alleles at different loci that are inherited together more frequently or less frequently than expected by their individual frequencies are said to show linkage disequilibrium.

Linkage Genetic linkage refers to the observation that two or more genes located on the same chromosome are inherited together. The ratio of being transmitted together versus being separated allows an estimate of their distance from each other (recombination fraction).

Locus The site on the chromosome where a gene is located.

Methylation (of DNA) The attachment of a methyl group to DNA. In vertebrates, this typically occurs at CpG sites (cytosine-phos-phate-guanine sites) in the DNA sequence, resulting in the conversion of cytosine to 5-methylcytosine.

Missense mutation A mutation that alters a codon so that it encodes a different amino acid.

mtDNa Mitochondria DNA.

Mutation The process that produces a gene or a chromosomal set differing from the wild type.

Nick translation A means of incorporating labelled nucleotides into a segment of nucleic acid by displacing random nicks (introduced by DNAase) through the use of DNA polymerase.

Nonsense codon A codon for which no normal tRNA molecule exists; the presence of a nonsense codon causes termination of translation (ending of the polypeptide chain). The three nonsense codons are called amber, ocher, and opal.

Nonsense mutation A mutation that alters a gene so as to produce a nonsense codon.

Northern blot RNA immobilized on a solid support after separation according to size by electrophoresis.

N-Terminus and C-Terminus The two ends of a protein/polypeptide chain. Protein sequences are given N?C in direction.

Nucleotide (nt) The structural unit of nucleic acid consisting of phosphate, sugar and purine or pyrimidine base.

Oligonucleotide A short (10-100 nt) stretch of ssDNA usually prepared by a series of chemical reaction following a known sequence.

Oncogene A type of gene associated with cancer production.

Operon A set of adjacent structural genes (prokaryotic) whose mRNA is synthesized in one piece, plus the adjacent regulatory signals that affect transcription of the structural genes. Regulation can be postive (turn it on when needed) or negative (keep it off until needed).

ORF (open reading frame) A section of a sequenced piece of DNA that begins with a start codon and ends with a stop codon; it is presumed to be the coding sequence of a gene.

Penetrance The frequency (in percent) with which a dominant or homozygous recessive gene or gene combination manifests itself in the phenotype of the carriers.

Pharmacogenetics A branch of genetics which deals with the genetic variability in individual responses to drugs and drug metabolism.

Phenocopy A nonhereditary, phenotypic modification (caused by special environmental conditions) that mimics a similar phenotype caused by a gene mutation.

Phenotype The observable properties (structural and functional) of an organism, produced by the interaction between the organism's genotype and the environment in which it finds itself.

Phenotype The form or expression taken by some character or gene in a specific individual. Genotype is not always reflected in the phenotype.

Plasmid A small, circular, extrachromosomal, self-replicating piece of DNA found in some bacteria (e.g. pBR322). Used in cloning (cloning plasmid = vector).

Pleiotropy Genes or mutations that result in the production of multiple, apparently unrelated, effects at the phenotypic level. For example, patients with phenylketonuria, caused by mutations in the *PAH* (phenylalanine hydroxylase) gene, have reduced hair and skin pigmentation in addition to mental retardation, resulting from toxic levels of phenylalanine.

Point mutation A mutation that can be mapped to one specific locus.

Poly(A)tail A string of adenine nucleotides added to mRNA after transcription.

Polycistronic mRNA An mRNA that encodes more than one protein (prokaryotes and viruses).

Polymerase chain reaction (PCR) Enzymatic technique to create multiple copies of one sequence of DNA

Polymorphism (genetic) A chromosome or DNA variant that is observed in natural populations. A gene locus is defined as polymorphic if a rare allele has a frequency of 0.01 (1%) or more.

Polymorphism The occurrence in a population of several phenotypic forms associated with alleles of one gene or homologs of one chromosome

Positional cloning Finding disease genes based on knowledge of their chromosomal location (usually found via linkage analysis in families with the disease) as opposed to knowledge of the function of the gene or protein encoded by the gene.

Primer A short single-stranded DNA or RNA that can act as a start site for 3' chain growth when bound to a single-stranded template.

Probe A fragment or sequence of ssDNA, dsDNA or RNA that will be hybridized to a complementary sequence of nucleotides in another single-strand nucleic acid (target). Probes are labeled in some what to make the reaction visible.

Promoter A regulator region a short distance from the 5' end (transcription start site) of a gene that acts as the binding site for RNA polymerase.

Reading frame The codon sequence that is determined by reading nucleotides in groups of three from some specific start codon.

Recombinant DNA The DNA molecule produced from inserting DNA from one organism into another piece of DNA by using genetic engineering techniques.

Regulatory genes Genes that have roles in turning on or off the transcription of other genes.

Replication fork The point at which the two strands of DNA are separated to allow replication of each strand.

Replication The process of making DNA.

Restriction enzyme Another name for restriction endonuclease.

Reverse Transcriptase An enzyme capable of synthesizing ssDNA from RNA. From retroviruses.

Ribosomal RNA (rRNA) A class of RNA molecules that have an integral role in ribosome structure and function.

RNA polymerase A DNA directed RNA polymerase that creates a strand of RNA complementary to one strand of the dsDNA template.

RNA Ribonucleic acid, the single stranded nucleic acid of three types; mRNA (messenger RNA), tRNA (transfer RNA) and rRNA (ribosomal RNA)

RNases Ubiquitous enzymes that degrade RNA

Second- or next-generation sequencing (also referred to as high-throughput sequencing) New techniques that have increased the speed and decreased the cost of DNA sequencing by two orders of magnitude, enabling the sequencing of the entire genomes of many individuals.

Sequencing A method that determines the actual sequence of the nucleotide bases in DNA or protein.

Signal sequence The N-terminal sequence of a secreted protein, which is required for transport through the cell membrane.

Single nucleotide polymorphism (SNP) Heritable polymorphism resulting from a single base pair change. SNPs generally have only two alleles.

Southern blot DNA that has been separated by gel electrophoresis, transferred from the gel to an immobile support (e.g. nitrocellulose or nylon), and bonded onto the support in single-stranded form ready for hybridization

ss Single stranded

Staggered cuts The cleavage of two opposite strands of duplex DNA at points near one another.

Streptavidin A bacterial form of avidin that has a slightly stronger affinity for biotin than does avidin form egg white.

Stringency The conditions of hybridization that increase the base-pairing of binding between two single strand portions of nucleic acid.

Structural variant Structural genomic variation includes any genetic variant that alters chromosomal structure, including inversions, translocations, duplications and deletions. Duplications and deletions, collectively known as CNVs (see *copy number variation)* are the most common form of structural variation in the human genome.

Supercoil A closed, double-stranded DNA molecule that is twisted on itself.

Synonymous nucleotide change/non-synonymous nucleotide change A change in the DNA sequence which does not result in the change in the amino acid sequence, eg, GTT>GTC both code for Valine (Val or V). A nonsynonymous change results in the coding of a different amino acid (eg, GTT>GAT results in Val>Asp).

Target nucleic acid DNA or RNA to be hybridized with the labelled probe.

Tissue-specific gene expression The expression of a gene in a specific and reproducible subset of tissues and cells during the development of a higher eukaryote.

Transcription The process of producing an RNA copy from a DNA template.

Translation The production of protein from messenger RNA.

Trinucleotide repeat expansion An increased number of contiguous trinucleotide repeats (eg, CAG, CGG) in the DNA sequence from one generation to the next. When the expansion extends into the pathological range, this type of mutation causes diseases such as Huntington's disease, fragile X syndrome, myotonic dystrophy, and many forms of spinocerebellar ataxia.

Uracil-N-glycosylase An enzyme that digests DNA that was replicated using dUTP instead of dTTP and is used to prevent amplicon carry over contamination.

Vector A plasmid, bacteriophage, cosmid or virus that carries foreign DNA into a host organism.

Western blot Proteins that have been separated by acrylamid electrophoresis, transferred and immobilized onto a solid support, then probed with labelled antibody.

X-inactivation The random, early embryological, inactivation of one of the X chromosomes in females, so that the expression of X-chromosomal genes is the same as that in males.

BIBLIOGRAPHY

Addo-Quaye, C., Eshoo, T. W., Bartel, D. P., and Axtell, M. J. (2008). Endogenous siRNA and miRNA targets identified by sequencing of the *Arabidopsis* degradome. *Curr. Biol.* 18, 758–762.

Aird, D., Ross, M. G., Chen, W. S., Danielsson, M., Fennell, T., Russ, C., Jaffe, D. B., Nusbaum, C., and Gnirke, A. (2011). Analyzing and minimizing PCR amplification bias in Illumina sequencing libraries. *Genome Biol.* 12, R18.

Albert, I., Mavrich, T. N., Tomsho, L. P., Qi, J., Zanton, S. J., Schuster, S. C., and Pugh, B. F. (2007). Translational and rotational settings of H2A.Z nucleosomes across the *Saccharomyces cerevisiae* genome. *Nature* 446, 572–576.

Ameur, A., Wetterbom, A., Feuk, L., and Gyllensten, U. (2010). Global and unbiased detection of splice junctions from RNA-seq data. *Genome Biol.* 11, R34.

Anders, S. (2010). *HTSeq: Analysing High-Throughput Sequencing Data With Python.* Available at: http://www-huber.embl.de/users/anders/HTSeq/doc/overview.html#author

Anders, S., and Huber, W. (2010). Differential expression analysis for sequence count data. *Genome Biol.* 11, R106.

Angeloni, F., Wagemaker, C. A., Jetten, M. S., Op Den Camp, H. J., Janssen-Megens, E. M., Francoijs, K. J., Stunnenberg, H. G., and Ouborg, N. J. (2011). De novo transcriptome characterization and development of genomic tools for *Scabiosa columbaria* L. using next-generation sequencing techniques. *Mol. Ecol. Resour.* 11, 662–674.

Arai-Kichise, Y., Shiwa, Y., Nagasaki, H., Ebana, K., Yoshikawa, H., Yano, M., and Wakasa, K. (2011). Discovery of genome-wide DNA polymorphisms in a landrace cultivar of Japonica rice by whole-genome sequencing. *Plant Cell Physiol.* 52, 274–282.

Armour, C. D., Castle, J. C., Chen, R., Babak, T., Loerch, P., Jackson, S., Shah, J. K., Dey, J., Rohl, C. A., Johnson, J. M., and Raymond, C. K. (2009). Digital transcriptome profiling using selective hexamer priming for cDNA synthesis. *Nat. Methods* 6, 647–649.

Au, K. F., Jiang, H., Lin, L., Xing, Y., and Wong, W. H. (2010). Detection of splice junctions from paired-end RNA-seq data by SpliceMap. *Nucleic Acids Res.* 38, 4570–4578.

Auer, P. L., and Doerge, R. W. (2010). Statistical design and analysis of RNA sequencing data. *Genetics* 185, 405–416.

Austin, R. S., Vidaurre, D., Stamatiou, G., Breit, R., Provart, N. J., Bonetta, D., Zhang, J., Fung, P., Gong, Y., Wang, P. W., Mccourt, P., and Guttman, D. S. (2011). Next-generation mapping of *Arabidopsis* genes. *Plant J.* 67, 715–725.

Bao, H., Guo, H., Wang, J., Zhou, R., Lu, X., and Shi, S. (2009). MapView: visualization of short reads alignment on a desktop computer. *Bioinformatics* 25, 1554–1555.

Bao, S., Jiang, R., Kwan, W., Wang, B., Ma, X., and Song, Y. Q. (2011). Evaluation of next-generation sequencing software in mapping and assembly. *J. Hum. Genet.* 56, 406–414.

Barski, A., Cuddapah, S., Cui, K., Roh, T. Y., Schones, D. E., Wang, Z., Wei, G., Chepelev, I., and Zhao, K. (2007). High-resolution profiling of histone methylations in the human genome. *Cell* 129, 823–837.

Barton, N. H. 1990. Pleiotropic models of quantitative variation. Genetics 124:773-782.

Barton, N. H., and M. Turelli. 1989. Evolutionary quantitative genetics: How little do we know? Annu. Rev. Genet. 23:337-370.

Bateson W (1902) Experiments with poultry. Rep Evol Comm Roy Soc 1: 87–124.

Beavis, W. D., D. Grant, M. Albertsen, and R. Fincher. 1991. Quantitative trait loci for plant height in four maize populations and their associations with qualitative genetic loci. Theor. Appl. Genet. 83:141-145.

Beckman, J. S., and M. Soller. 1986. Restriction fragment length polymorphisms and genetic improvement of agricultural species. Euphytica 35:111-124.

Bernardo, R. 1992. Relationship between single-cross performance and molecular marker heterozygosity. Theor. Appl. Genet. 83:628-634.

Blankenberg, D., Gordon, A., Von Kuster, G., Coraor, N., Taylor, J., and Nekrutenko, A. (2010). Manipulation of FASTQ data with galaxy. *Bioinformatics* 26, 1783–1785.

Blow, N. (2009). Genomics: catch me if you can. *Nat. Methods* 6, 539–544.

Boyle, A. P., Davis, S., Shulha, H. P., Meltzer, P., Margulies, E. H., Weng, Z., Furey, T. S., and Crawford, G. E. (2008). High-resolution mapping and characterization of open chromatin across the genome. *Cell* 132, 311–322.

Bradford, J. R., Hey, Y., Yates, T., Li, Y., Pepper, S. D., and Miller, C. J. (2010). A comparison of massively parallel nucleotide sequencing with oligonucleotide microarrays for global transcription profiling. *BMC Genomics* 11, 282. doi:10.1186/1471-2164-11-282

Bullard, J. H., Mostovoy, Y., Dudoit, S., and Brem, R. B. (2010a). Polygenic and directional regulatory evolution across pathways in *Saccharomyces*. *Proc. Natl. Acad. Sci. U.S.A.* 107, 5058–5063.

Bullard, J. H., Purdom, E., Hansen, K. D., and Dudoit, S. (2010b). Evaluation of statistical methods for normalization and differential expression in mRNA-seq experiments. *BMC Bioinformatics* 11, 94. doi:10.1186/1471-2105-11-94

Burr, B., S. V. Evola, F. A. Burr, and J. S. Beckman. 1983. The application of restriction fragment length polymorphisms to plant breeding. p. 45-59. In: J. K. Setlow and A. Hollander (eds.) Genetic engineering: Principles and methods, Vol. 5. Plenum Publishing Corp., New York, NY.

Carlson, M., Pages, H., Aboyoun, P., Falcon, S., Morgan, M., Sarkar, D., and Lawrence, M. (2009). *Genomic Features: Tools for Making and Manipulating Transcript Centric Annotations.*

Chen, K., Wallis, J. W., Mclellan, M. D., Larson, D. E., Kalicki, J. M., Pohl, C. S., Mcgrath, S. D., Wendl, M. C., Zhang, Q., Locke, D. P., Shi, X., Fulton, R. S., Ley, T. J., Wilson, R. K., Ding, L., and Mardis, E. R. (2009). BreakDancer: an algorithm for high-resolution mapping of genomic structural variation. *Nat. Methods* 6, 677–681.

Christodoulou, D. C., Gorham, J. M., Herman, D. S., and Seidman, J. (2011). Construction of normalized RNA-seq libraries for next-generation sequencing using the crab duplex-specific nuclease. *Curr. Protoc. Mol. Biol.* 94, 4.12.1–4.12.11.

Churchman, L. S., and Weissman, J. S. (2011). Nascent transcript sequencing visualizes transcription at nucleotide resolution. *Nature* 469, 368–373.

Cock, P. J., Antao, T., Chang, J. T., Chapman, B. A., Cox, C. J., Dalke, A., Friedberg, I., Hamelryck, T., Kauff, F., Wilczynski, B., and De Hoon, M. J. (2009). Biopython: freely available Python tools for computational molecular biology and bioinformatics. *Bioinformatics* 25, 1422–1423.

Coe, E. H. Jr., M. G. Neuffer, and D. A. Hoisington. 1988. The genetics of corn. p. 83-258. In: G. F. Sprague and J. W. Dudley (eds.) Corn and Corn Improvement. Am. Soc. Agron., Madison, WI. Cowen, N. M. 1988. The use of replicated progenies in marker-based mapping of QTL's. Theor. Appl. Genet. 75:857-865.

Cokus, S. J., Feng, S., Zhang, X., Chen, Z., Merriman, B., Haudenschild, C. D., Pradhan, S., Nelson, S. F., Pellegrini, M., and Jacobsen, S. E. (2008). Shotgun bisulphite sequencing of the *Arabidopsis* genome reveals DNA methylation patterning. *Nature* 452, 215–219.

Core, L. J., Waterfall, J. J., and Lis, J. T. (2008). Nascent RNA sequencing reveals widespread pausing and divergent initiation at human promoters. *Science* 322, 1845–1848.

Costa, V., Angelini, C., De Feis, I., and Ciccodicola, A. (2010). Uncovering the complexity of transcriptomes with RNA-seq. *J. Biomed. Biotechnol.* 2010, 853916.

Cowles, C. R., Hirschhorn, J. N., Altshuler, D., and Lander, E. S. (2002). Detection of regulatory variation in mouse genes. *Nat. Genet.* 32, 432–437.

Cox, T. S. 1992. Use of molecular markers for analyzing quantitative traits in wheat. In: Proc. 19th Hard Red Winter Wheat Workers Conf.

Crossa, J. 1990. Statistical analyses of multilocation trials. Adv. Agron. 44:55-85.

Darlington, C. D. 1937. The biology of crossing-over. Nature 140:759-761.

Decook, R., Lall, S., Nettleton, D., and Howell, S. H. (2006). Genetic regulation of gene expression during shoot development in *Arabidopsis*. *Genetics* 172, 1155–1164.

Degner, J. F., Marioni, J. C., Pai, A. A., Pickrell, J. K., Nkadori, E., Gilad, Y., and Pritchard, J. K. (2009). Effect of read-mapping biases on detecting allele-specific expression from RNA-sequencing data. *Bioinformatics* 25, 3207–3212.

Depristo, M. A., Banks, E., Poplin, R., Garimella, K. V., Maguire, J. R., Hartl, C., Philippakis, A. A., Del Angel, G., Rivas, M. A., Hanna, M., Mckenna, A., Fennell, T. J., Kernytsky, A. M., Sivachenko, A. Y., Cibulskis, K., Gabriel, S. B., Altshuler, D., and Daly, M. J. (2011). A framework for variation discovery and genotyping using next-generation DNA sequencing data. *Nat. Genet.* 43, 491–498.

Dohm, J. C., Lottaz, C., Borodina, T., and Himmelbauer, H. (2008). Substantial biases in ultra-short read data sets from high-throughput DNA sequencing. *Nucleic Acids Res.* 36, e105.

Dorshorst B, Harun-Or-Rashid M, Bagherpoor AJ, Rubin CJ, Ashwell C, et al. (2015) A genomic duplication is associated with ectopic Eomesodermin expression in the embryonic chicken comb and two duplex-comb phenotypes. PLoS Genet 11(1): e1004947.

Down, T. A., Rakyan, V. K., Turner, D. J., Flicek, P., Li, H., Kulesha, E., Graf, S., Johnson, N., Herrero, J., Tomazou, E. M., Thorne, N. P., Backdahl, L., Herberth, M., Howe, K. L., Jackson, D. K., Miretti, M. M., Marioni, J. C., Birney, E., Hubbard, T. J., Durbin, R., Tavare, S., and Beck, S. (2008). A Bayesian deconvolution strategy for immunoprecipitation-based DNA methylome analysis. *Nat. Biotechnol.* 26, 779–785.

Drost, D. R., Benedict, C. I., Berg, A., Novaes, E., Novaes, C. R., Yu, Q., Dervinis, C., Maia, J. M., Yap, J., Miles, B., and Kirst, M. (2010). Diversification in the genetic architecture of gene expression and transcriptional networks in organ differentiation of *Populus*. *Proc. Natl. Acad. Sci. U.S.A.* 107, 8492–8497.

Emerson, K. J., Merz, C. R., Catchen, J. M., Hohenlohe, P. A., Cresko, W. A., Bradshaw, W. E., and Holzapfel, C. M. (2010b). Resolving postglacial phylogeography using high-throughput sequencing. *Proc. Natl. Acad. Sci. U.S.A.* 107, 16196–16200.

ENCODE. (2011). Standards, guidelines and best practices for RNA-seq. V1.0.

Etter, P. D., Preston, J. L., Bassham, S., Cresko, W. A., and Johnson, E. A. (2011). Local de novo assembly of RAD paired-end contigs using short sequencing reads. *PLoS ONE* 6, e18561. doi:10.1371/journal.pone.0018561

Everett, M. V., Grau, E. D., and Seeb, J. E. (2011). Short reads and nonmodel species: exploring the complexities of next-generation sequence assembly and SNP discovery in the absence of a reference genome. *Mol. Ecol. Resour.* 11(Suppl. 1), 93–108.

Falconer, D. S. 1981. Introduction to quantitative genetics. 2nd ed. Longman, New York.

Falgueras, J., Lara, A. J., Fernandez-Pozo, N., Canton, F. R., Perez-Trabado, G., and Claros, M. G. (2010). SeqTrim: a high-throughput pipeline for pre-processing any type of sequence read. *BMC Bioinformatics* 11, 38. doi:10.1186/1471-2105-11-38

Fang, Z., and Cui, X. (2011). Design and validation issues in RNA-seq experiments. *Brief. Bioinform.* 12, 280–287.

Faulkner, G. J., Forrest, A. R., Chalk, A. M., Schroder, K., Hayashizaki, Y., Carninci, P., Hume, D. A., and Grimmond, S. M. (2008). A rescue strategy for multi mapping short sequence tags refines surveys of transcriptional activity by CAGE. *Genomics* 91, 281–288.

Fehr, W. R. (ed.) 1984. Genetic contributions to yield gains of five major crop plants. Special publ. No. 7. Crop Sci. Soc. of Am., Madison, WI.

Fehr, W. R. (ed.). 1987. Principles of cultivar development. Vol. 2. Crop Species. Macmillan Publ. Co., New York.

Fisher, R. A. (1918). The correlation between relatives on the supposition of Mendelian inheritance. *Philos. Trans. R. Soc. Edinb.* 52, 399–433.

Fisher, R. A. 1918. The correlation between relatives under the supposition of mendelian inheritance. Trans. R. Soc. Edinburgh 52:399-433.

Fontanillas, P., Landry, C. R., Wittkopp, P. J., Russ, C., Gruber, J. D., Nusbaum, C., and Hartl, D. L. (2010). Key considerations for measuring allelic expression on a genomic scale using high-throughput sequencing. *Mol. Ecol.* 19(Suppl. 1), 212–227.

Frohler, S., and Dieterich, C. (2010). ACCUSA – accurate SNP calling on draft genomes. *Bioinformatics* 26, 1364–1365.

Garber, M., Grabherr, M. G., Guttman, M., and Trapnell, C. (2011). Computational methods for transcriptome annotation and quantification using RNA-seq. *Nat. Methods* 8, 469–477.

Gardner, E. J., Simmons, M. J., & Snustad, D. P. (1991). Principles of genetics. New York: J. Wiley.

Garg, R., Patel, R. K., Jhanwar, S., Priya, P., Bhattacharjee, A., Yadav, G., Bhatia, S., Chattopadhyay, D., Tyagi, A. K., and Jain, M. (2011). Gene discovery and tissue-specific transcriptome analysis in chickpea with massively parallel pyrosequencing and web resource development. *Plant Physiol.* 156, 1661–1678.

Geraldes, A., Pang, J., Thiessen, N., Cezard, T., Moore, R., Zhao, Y., Tam, A., Wang, S., Friedmann, M., Birol, I., Jones, S. J., Cronk, Q. C., and Douglas, C. J. (2011). SNP discovery in black cottonwood (*Populus trichocarpa*) by population transcriptome resequencing. *Mol. Ecol. Resour.* 11(Suppl. 1), 81–92.

Glenn, T. C. (2011). Field guide to next-generation DNA sequencers. *Mol. Ecol. Resour.* 11, 759–769.

Goecks, J., Nekrutenko, A., and Taylor, J. (2010). Galaxy: a comprehensive approach for supporting accessible, reproducible, and transparent computational research in the life sciences. *Genome Biol.* 11, R86.

Haag-Liautard, C., *et al.* Direct estimation of per nucleotide and genomic deleterious mutation rates in *Drosophila*. *Nature* **445**, 82–85 (2007).

Haldane, J. B. S. 1932. The causes of evolution. Longmans, Green, London.

Hallauer, A. R. 1990. Methods used in developing maize inbreds. Maydica 35:1-16.

Hallauer, A. R. and J. B. Miranda, FO. 1988. Quantitative genetics in maize breeding. 2nd ed. Iowa State University Press, Ames.

Hansen, K. D., Brenner, S. E., and Dudoit, S. (2010). Biases in Illumina transcriptome sequencing caused by random hexamer priming. *Nucleic Acids Res.* 38, e131.

Hansen, K. D., Zhijin, W., Irizarry, R. A., and Leek, J. T. (2011). Sequencing technology does not eliminate biological variability. *Nat. Biotechnol.* 29, 575–573.

Hardcastle, T. J., and Kelly, K. A. (2010). baySeq: empirical Bayesian methods for identifying differential expression in sequence count data. *BMC Bioinformatics* 11, 422. doi:10.1186/1471-2105-11-422

Hashimoto, T., De Hoon, M. J., Grimmond, S. M., Daub, C. O., Hayashizaki, Y., and Faulkner, G. J. (2009). Probabilistic resolution of multi-mapping reads in massively parallel sequencing data using MuMRescueLite. *Bioinformatics* 25, 2613–2614.

Hawkins, R. D., Hon, G. C., and Ren, B. (2010). Next-generation genomics: an integrative approach. *Nat. Rev. Genet.* 11, 476–486.

He, Y., Vogelstein, B., Velculescu, V. E., Papadopoulos, N., and Kinzler, K. W. (2008). The antisense transcriptomes of human cells. *Science* 322, 1855–1857.

Helentjaris, T. G. 1992. RFLP analyses for manipulating agronomic traits in plants. p. 357-372. In: H. R. Stalker and J. P. Murphy (eds.) Plant breeding in the 1990s. CAB International, Wallingford, UK.

Helentjaris, T. G., G. King, M. Slocum, C. Siedenstrang, and S. Wegman. 1985. Restriction fragment polymorphisms as probes for plant diversity and their development as tools for applied plant breeding. Plant Mol. Biol. 5:109-118.

Hoekstra, H. E., and Coyne, J. A. (2007). The locus of evolution: evo devo and the genetics of adaptation. *Evolution* 61, 995–1016.

Hohenlohe, P. A., Amish, S. J., Catchen, J. M., Allendorf, F. W., and Luikart, G. (2011). Next-generation RAD sequencing identifies thousands of SNPs for assessing hybridization between rainbow and west slope cutthroat trout. *Mol. Ecol. Resour.* 11(Suppl. 1), 117–122.

Hong, L. Z., Li, J., Schmidt-Kuntzel, A., Warren, W. C., and Barsh, G. S. (2011). Digital gene expression for non-model organisms. *Genome Res.* doi: 10.1101/gr.122135.111. [Epub ahead of print].

Hormozdiari, F., Alkan, C., Eichler, E. E., and Sahinalp, S. C. (2009). Combinatorial algorithms for structural variation detection in high-throughput sequenced genomes. *Genome Res.* 19, 1270–1278.

Hormozdiari, F., Hajirasouliha, I., Dao, P., Hach, F., Yorukoglu, D., Alkan, C., Eichler, E. E., and Sahinalp, S. C. (2010). Next-generation VariationHunter: combinatorial algorithms for transposon insertion discovery. *Bioinformatics* 26, i350–i357.

Horner, E. S., E. Magloire, and J. A. Morera. 1989. Comparison of selection for S2 progeny vs. testcross performance for population improvement in maize. Crop Sci. 29:868-874.

http://kmbiology.weebly.com/mendel-and-genetics—notes.html

http://knowgenetics.org/mendelian-genetics/

https://en.wikipedia.org/wiki/Mendelian_inheritance

https://www.cliffsnotes.com/study-guides/biology/plant-biology/genetics/mendelian-genetics

Huang, W., and Marth, G. (2008). EagleView: a genome assembly viewer for next-generation sequencing technologies. *Genome Res.* 18, 1538–1543.

Ingolia, N. T. (2010). Genome-wide translational profiling by ribosome footprinting. *Meth. Enzymol.* 470, 119–142.

Ingolia, N. T., Ghaemmaghami, S., Newman, J. R., and Weissman, J. S. (2009). Genome-wide analysis in vivo of translation with nucleotide resolution using ribosome profiling. *Science* 324, 218–223.

Johnson, D. S., Mortazavi, A., Myers, R. M., and Wold, B. (2007). Genome-wide mapping of in vivo protein-DNA interactions. *Science* 316, 1497–1502.

Jones FC, Grabherr MG, Chan YF, Russell P, Mauceli E, et al. (2012) The genomic basis of adaptive evolution in threespine sticklebacks. Nature 484: 55–61. pmid:22481358

Katz, Y., Wang, E. T., Airoldi, E. M., and Burge, C. B. (2010). Analysis and design of RNA sequencing experiments for identifying isoform regulation. *Nat. Methods* 7, 1009–1015.

Kaur, S., Cogan, N. O., Pembleton, L. W., Shinozuka, M., Savin, K. W., Materne, M., and Forster, J. W. (2011). Transcriptome sequencing of lentil based on second-generation technology permits large-scale unigene assembly and SSR marker discovery. *BMC Genomics* 12, 265. doi:10.1186/1471-2164-12-265

Kempthorne, and T. B. Bailey, Jr. (eds.) Proc. of the Int. Conf. on Quantitative Genetics, 16-21 Aug. 1976. Iowa State University Press, Ames.

Kempthorne, O. 1988. An overview of the field of quantitative genetics. p. 47-56. In: B. S. Weir, E. J. Eisen, M. M. Goodman, and G. Namkoong (eds.) Proc. of the Second Int. Conf. on Quantitative Genetics, 31 May - 5 June 1987, Sinauer Associates Inc., Sunderland, Mass.

Kenny, E. M., Cormican, P., Gilks, W. P., Gates, A. S., O'dushlaine, C. T., Pinto, C., Corvin, A. P., Gill, M., and Morris, D. W. (2011). Multiplex target enrichment using DNA indexing for ultra-high throughput SNP detection. *DNA Res.* 18, 31–38.

Kerstens, H., Crooijmans, R., Veenendaal, A., Dibbits, B., Chin-a-Woeng, T., Den Dunnen, J., and Groenen, M. (2009). Large scale single nucleotide polymorphism discovery in unsequenced genomes using second generation high throughput sequencing technology: applied to turkey. *BMC Genomics* 10, 479. doi:10.1186/1471-2164-10-479

Kertesz, M., Wan, Y., Mazor, E., Rinn, J. L., Nutter, R. C., Chang, H. Y., and Segal, E. (2010). Genome-wide measurement of RNA secondary structure in yeast. *Nature* 467, 103–107.

Keurentjes, J. J. B., Fu, J., Terpstra, I. R., Garcia, J. M., Van Den Ackerveken, G., Snoek, L. B., Peeters, A. J. M., Vreugdenhil, D., Koornneef, M., and Jansen, R. C. (2007). Regulatory network construction in *Arabidopsis* by using genome-wide gene expression quantitative trait loci. *Proc. Natl. Acad. Sci. U.S.A.* 104, 1708–1713.

Kirst, M., Basten, C. J., Myburg, A. A., Zeng, Z. B., and Sederoff, R. R. (2005). Genetic architecture of transcript-level variation in differentiating xylem of a eucalyptus hybrid. *Genetics* 169, 2295–2303.

Kliebenstein, D. J., West, M. A., Van Leeuwen, H., Loudet, O., Doerge, R. W., and St Clair, D. A. (2006). Identification of QTLs controlling gene expression networks defined a priori. *BMC Bioinformatics* 7, 308. doi:10.1186/1471-2105-7-308

Knapp, S. J. 1989. Quasi-Mendelian analyses of quantitative traits using molecular marker linkage maps. p. 51-67. In: G. Robbelen (ed.) Science for plant breeding, Proceedings of the XII congress of EUCARPIA. Paul Parey Scientific Publishers, Berlin.

Lalonde, E., Ha, K. C., Wang, Z., Bemmo, A., Kleinman, C. L., Kwan, T., Pastinen, T., and Majewski, J. (2011). RNA sequencing reveals the role of splicing polymorphisms in regulating human gene expression. *Genome Res.* 21, 545–554.

Lande, R. 1981. The minimum number of genes contributing to quantitative variation between and within populations. Genetics 99:541-553.

Lande, R. 1992. Marker-assisted selection in relation to traditional methods of plant breeding. p. 437-451. In: H. R. Stalker and J. P. Murphy (eds.) Plant breeding in the 1990s. CAB International, Wallingford, UK.

Lande, R. and R. Thompson. 1990. Efficiency of marker-assisted selection in the improvement of quantitative traits. Genetics 124:743-756.

Lander, E. S. and D. Botstein. 1989. Mapping mendelian factors underlying quantitative traits using RFLP linkage maps. Genetics 124:743-756.

Liu, S., Lin, L., Jiang, P., Wang, D., and Xing, Y. (2010). A comparison of RNA-seq and high-density exon array for detecting differential gene expression between closely related species. *Nucleic Acids Res.* 39, 578–588.

Loewe, L., & Charlesworth, B. Inferring the distribution of mutational effects on fitness in *Drosophila*. *Biology Letters* **2**, 426–430 (2006)

Mather, K. 1941. Variation and selection of polygenic characters. J. Genet. 41:159-193.

Matsumura, H., Yoshida, K., Luo, S., Kimura, E., Fujibe, T., Albertyn, Z., Barrero, R. A., Kruger, D. H., Kahl, G., Schroth, G. P., and Terauchi, R. (2010). High-throughput SuperSAGE for digital gene expression analysis of multiple samples using next generation sequencing. *PLoS ONE* 5, e12010. doi:10.1371/journal.pone.0012010

Mcintyre, L. M., Lopiano, K. K., Morse, A. M., Amin, V., Oberg, A. L., Young, L. J., and Nuzhdin, S. V. (2011). RNA-seq: technical variability and sampling. *BMC Genomics* 12, 293. doi:10.1186/1471-2164-12-293

Mcmanus, C. J., Coolon, J. D., Duff, M. O., Eipper-Mains, J., Graveley, B. R., and Wittkopp, P. J. (2010). Regulatory divergence in *Drosophila* revealed by mRNA-seq. *Genome Res.* 20, 816–825.

Melchinger, A. E. 1990. Use of molecular markers in breeding for oligogenic disease resistance. Plant Breeding 104:1-19.

Melchinger, A.E. 1993. Use of RFLP markers for analyses of genetic relationships among breeding materials and prediction of hybrid performance. Proc. of the First Int. Crop Sci. Cong.

Melchinger, A.E., M. Messmer, M. Lee, W. L. Woodman, and K. R. Lamkey. 1991. Diversity and relationships among U.S. maize inbreds revealed by restriction fragment length polymorphisms. Crop Sci. 31:669-678.

Metzker, M. L. (2010). Sequencing technologies – the next generation. *Nat. Rev. Genet.* 11, 31–46.

Milne, I., Bayer, M., Cardle, L., Shaw, P., Stephen, G., Wright, F., and Marshall, D. (2010). Tablet – next generation sequence assembly visualization. *Bioinformatics* 26, 401–402.

Mizrachi, E., Hefer, C. A., Ranik, M., Joubert, F., and Myburg, A. A. (2010). De novo assembled expressed gene catalog of a fast-growing *Eucalyptus* tree produced by Illumina mRNA-seq. *BMC Genomics* 11, 681.

Montes, J. M., Melchinger, A. E., and Reif, J. C. (2007). Novel throughput phenotyping platforms in plant genetic studies. *Trends Plant Sci.* 12, 433–436.

Montgomery, S. B., Sammeth, M., Gutierrez-Arcelus, M., Lach, R. P., Ingle, C., Nisbett, J., Guigo, R., and Dermitzakis, E. T. (2010). Transcriptome genetics using second generation sequencing in a Caucasian population. *Nature* 464, 773–777.

Ness, R. W., Siol, M., and Barrett, S. C. (2011). De novo sequence assembly and characterization of the floral transcriptome in cross- and self-fertilizing plants. *BMC Genomics* 12, 298. doi:10.1186/1471-2164-12-298

Nicolae, M., Mangul, S., Mandoiu, I. I., and Zelikovsky, A. (2011). Estimation of alternative splicing isoform frequencies from RNA-seq data. *Algorithms Mol. Biol.* 6, 9.

Niedringhaus, T. P., Milanova, D., Kerby, M. B., Snyder, M. P., and Barron, A. E. (2011). Landscape of next-generation sequencing technologies. *Anal. Chem.* 83, 4327–4341.

Nielsen, R., Paul, J. S., Albrechtsen, A., and Song, Y. S. (2011). Genotype and SNP calling from next-generation sequencing data. *Nat. Rev. Genet.* 12, 443–451.

Nijman, I. J., Mokry, M., Van Boxtel, R., Toonen, P., De Bruijn, E., and Cuppen, E. (2010). Mutation discovery by targeted genomic enrichment of multiplexed barcoded samples. *Nat. Methods* 7, 913–915.

Orr, H. A. The genetic theory of adaptation: A brief history. *Nature Review Genetics* **6**, 119–127 (2005) doi:10.1038/nrg1523 (link to article)

Oshlack, A., and Wakefield, M. J. (2009). Transcript length bias in RNA-seq data confounds systems biology. *Biol. Direct* 4, 14.

Ossowski, S., Schneeberger, K., Clark, R. M., Lanz, C., Warthmann, N., and Weigel, D. (2008). Sequencing of natural strains of *Arabidopsis thaliana* with short reads. *Genome Res.* 18, 2024–2033.

Parchman, T. L., Geist, K. S., Grahnen, J. A., Benkman, C. W., and Buerkle, C. A. (2010). Transcriptome sequencing in an ecologically important tree species: assembly, annotation, and marker discovery. *BMC Genomics* 11, 180. doi:10.1186/1471-2164-11-180

Parkhomchuk, D., Borodina, T., Amstislavskiy, V., Banaru, M., Hallen, L., Krobitsch, S., Lehrach, H., and Soldatov, A. (2009). Transcriptome analysis by strand-specific sequencing of complementary DNA. *Nucleic Acids Res.* 37, e123.

Potokina, E., Druka, A., Luo, Z., Wise, R., Waugh, R., and Kearsey, M. (2008). Gene expression quantitative trait locus analysis of 16 000 barley genes reveals a complex pattern of genome-wide transcriptional regulation. *Plant J.* 53, 90–101.

Quince, C., Lanzen, A., Davenport, R. J., and Turnbaugh, P. J. (2011). Removing noise from pyrosequenced amplicons. *BMC Bioinformatics* 12, 38. doi:10.1186/1471-2105-12-38

Quinlan, A. R., and Hall, I. M. (2010). BEDTools: a flexible suite of utilities for comparing genomic features. *Bioinformatics* 26, 841–842.

Ratan, A., Zhang, Y., Hayes, V., Schuster, S., and Miller, W. (2010). Calling SNPs without a reference sequence. *BMC Bioinformatics* 11, 130. doi:10.1186/1471-2105-11-130

Raz, T., Kapranov, P., Lipson, D., Letovsky, S., Milos, P. M., and Thompson, J. F. (2011). Protocol dependence of sequencing-based gene expression measurements. *PLoS ONE* 6, e19287. doi:10.1371/journal.pone.0019287

Robertson, G., Hirst, M., Bainbridge, M., Bilenky, M., Zhao, Y., Zeng, T., Euskirchen, G., Bernier, B., Varhol, R., Delaney, A., Thiessen, N., Griffith, O. L., He, A., Marra, M., Snyder, M., and Jones, S. (2007). Genome-wide profiles of STAT1 DNA association using chromatin immunoprecipitation and massively parallel sequencing. *Nat. Methods* 4, 651–657.

Robinson, J. T., Thorvaldsdottir, H., Winckler, W., Guttman, M., Lander, E. S., Getz, G., and Mesirov, J. P. (2011). Integrative genomics viewer. *Nat. Biotechnol.* 29, 24–26.

Sandelin, A., *et al.* Arrays of ultraconserved non-coding regions span the loci of key developmental genes in vertebrate genomes. *BMC Genomics* **5**, 99 (2004)

Schlotterer, C. (2004). The evolution of molecular markers – just a matter of fashion? *Nat. Rev. Genet.* 5, 63–69.

Schmieder, R., and Edwards, R. (2011). Quality control and preprocessing of metagenomic datasets. *Bioinformatics* 27, 863–864.

Schmieder, R., Lim, Y. W., Rohwer, F., and Edwards, R. (2010). TagCleaner: identification and removal of tag sequences from genomic and metagenomic datasets. *BMC Bioinformatics* 11, 341. doi:10.1186/1471-2105-11-341

Schneeberger, K., and Weigel, D. (2011). Fast-forward genetics enabled by new sequencing technologies. *Trends Plant Sci.* 16, 282–288.

Schneeberger, K., Ossowski, S., Lanz, C., Juul, T., Petersen, A. H., Nielsen, K. L., Jorgensen, J. E., Weigel, D., and Andersen, S. U. (2009). SHOREmap: simultaneous mapping and mutation identification by deep sequencing. *Nat. Methods* 6, 550–551.

Schwartz, S., Oren, R., and Ast, G. (2011). Detection and removal of biases in the analysis of next-generation sequencing reads. *PLoS ONE* 6, e16685. doi:10.1371/journal.pone.0016685

Skelly, D. A., Johansson, M., Madeoy, J., Wakefield, J., and Akey, J. M. (2011). A powerful and flexible statistical framework for testing hypotheses of allele-specific gene expression from RNA-seq data. *Genome Res.* doi: 10.1101/gr.119784.110. [Epub ahead of print].

Souaiaia, T., Frazier, Z., and Chen, T. (2011). ComB: SNP calling and mapping analysis for color and nucleotide space platforms. *J. Comput. Biol.* 18, 795–807.

Stein, L. D. (2010). The case for cloud computing in genome informatics. *Genome Biol.* 11, 207.

Stern, D. L., and Orgogozo, V. (2008). The loci of evolution: how predictable is genetic evolution? *Evolution* 62, 2155–2177.

Tarazona, S., Garcia-Alcalde, F., Dopazo, J., Ferrer, A., and Conesa, A. (2011). Differential expression in RNA-seq: a matter of depth. *Genome Res.* doi: 10.1101/gr.124321.111. [Epub ahead of print].

Toung, J. M., Morley, M., Li, M., and Cheung, V. G. (2011). RNA-sequence analysis of human B-cells. *Genome Res.* 21, 991–998.

Trapnell, C., Pachter, L., and Salzberg, S. L. (2009). TopHat: discovering splice junctions with RNA-seq. *Bioinformatics* 25, 1105–1111.

Trapnell, C., Williams, B. A., Pertea, G., Mortazavi, A., Kwan, G., Van Baren, M. J., Salzberg, S. L., Wold, B. J., and Pachter, L. (2010). Transcript assembly and quantification by RNA-seq reveals unannotated transcripts and isoform switching during cell differentiation. *Nat. Biotechnol.* 28, 511–515.

Trick, M., Long, Y., Meng, J., and Bancroft, I. (2009). Single nucleotide polymorphism (SNP) discovery in the polyploid *Brassica napus* using Solexa transcriptome sequencing. *Plant Biotechnol. J.* 7, 334–346.

Underwood, J. G., Uzilov, A. V., Katzman, S., Onodera, C. S., Mainzer, J. E., Mathews, D. H., Lowe, T. M., Salama, S. R., and Haussler, D. (2010). FragSeq: transcriptome-wide RNA structure probing using high-throughput sequencing. *Nat. Methods* 7, 995–1001.

Van Tassell, C. P., Smith, T. P., Matukumalli, L. K., Taylor, J. F., Schnabel, R. D., Lawley, C. T., Haudenschild, C. D., Moore, S. S., Warren, W. C., and Sonstegard, T. S. (2008). SNP discovery and allele frequency estimation by deep sequencing of reduced representation libraries. *Nat. Methods* 5, 247–252.

Verma, P. S., & Agrawal, V. K. (2006). Cell Biology, Genetics, Molecular Biology, Evolution & Ecology (1 ed.). S .Chand and company Ltd.

Vuylsteke, M., Daele, H., Vercauteren, A., Zabeau, M., and Kuiper, M. (2006). Genetic dissection of transcriptional regulation by cDNA-AFLP. *Plant J.* 45, 439–446.

Vuylsteke, M., Van Eeuwijk, F., Van Hummelen, P., Kuiper, M., and Zabeau, M. (2005). Genetic analysis of variation in gene expression in *Arabidopsis thaliana. Genetics* 171, 1267–1275.

Wang, J., Huda, A., Lunyak, V. V., and Jordan, I. K. (2010a). A Gibbs sampling strategy applied to the mapping of ambiguous short-sequence tags. *Bioinformatics* 26, 2501–2508.

Wang, K., Singh, D., Zeng, Z., Coleman, S. J., Huang, Y., Savich, G. L., He, X., Mieczkowski, P., Grimm, S. A., Perou, C. M., Macleod, J. N., Chiang, D. Y., Prins, J. F., and Liu, J. (2010b). MapSplice: accurate mapping of RNA-seq reads for splice junction discovery. *Nucleic Acids Res.* 38, e178.

Wang, L., Feng, Z., Wang, X., and Zhang, X. (2010c). DEGseq: an R package for identifying differentially expressed genes from RNA-seq data. *Bioinformatics* 26, 136–138.

Wang, X., Wu, Z., and Zhang, X. (2010d). Isoform abundance inference provides a more accurate estimation of gene expression levels in RNA-seq. *J. Bioinform. Comput. Biol.* 8(Suppl. 1), 177–192.

Wang, Z., Fang, B., Chen, J., Zhang, X., Luo, Z., Huang, L., Chen, X., and Li, Y. (2010e). De novo assembly and characterization of root transcriptome using Illumina paired-end sequencing and development of cSSR markers in sweet potato (*Ipomoea batatas*). *BMC Genomics* 11, 726. doi:10.1186/1471-2164-11-726

Wang, Z., Gerstein, M., and Snyder, M. (2009). RNA-seq: a revolutionary tool for transcriptomics. *Nat. Rev. Genet.* 10, 57–63.

Wei, W., Qi, X., Wang, L., Zhang, Y., Hua, W., Li, D., Lv, H., and Zhang, X. (2011). Characterization of the sesame (*Sesamum indicum* L.) global transcriptome using Illumina paired-end sequencing and development of EST-SSR markers. *BMC Genomics* 12, 451. doi:10.1186/1471-2164-12-451

Weir, B. S. 1990. Genetic data analysis. Sinauer Associates, Inc., Suderland, Massachusetts.

Weir, B. S. and C. C. Cockerham. 1977. Two-locus theory in quantitative genetics. p. 247-269. In: E. Pollak, O. Kempthorne, and T. B. Bailey, Jr. (eds.) Proc. of the Int. Conf. on Quantitative Genetics, 16-21 Aug. 1976. Iowa State University Press, Ames.

Weir, B. S., E. J. Eisen, M. M. Goodman, and G. Namkoong (eds.). 1988. Proceedings of the second international conference on quantitative genetics. Sinauer Assoc. Inc., Sunderland, MA.

Weller, J. I. 1987. Mapping and analysis of quantitative trait loci in Lycopersicon (tomato) with the aid of genetic markers using approximate maximum likelihood methods. Heredity 39:413-421.

West, M. A. L., Kim, K., Kliebenstein, D. J., Van Leeuwen, H., Michelmore, R. W., Doerge, R. W., and St Clair, D. A. (2007). Global eQTL mapping reveals the complex genetic architecture of transcript-level variation in *Arabidopsis*. *Genetics* 175, 1441–1450.

Willing, E. M., Hoffmann, M., Klein, J. D., Weigel, D., and Dreyer, C. (2011). Paired-end RAD-seq for de novo assembly and marker design without available reference. *Bioinformatics* 27, 2187–2193.

Wittkopp PJ, Kalay G (2012) Cis-regulatory elements: molecular mechanisms and evolutionary processes underlying divergence. Nat Rev Genet 13: 59–69. pmid:22143240

Wittkopp, P. J., Haerum, B. K., and Clark, A. G. (2004). Evolutionary changes in cis and trans gene regulation. *Nature* 430, 85–88.

Wray, G. A. (2007). The evolutionary significance of cis-regulatory mutations. *Nat. Rev. Genet.* 8, 206–216.

Wright D, Boije H, Meadows JR, Bed'hom B, Gourichon D, et al. (2009) Copy number variation in intron 1 of SOX5 causes the Pea-comb phenotype in chickens.

Wright, S. 1968. Evolution and genetics of populations. Vol. I. Genetic and biometric foundations. University of Chicago Press, Chicago.

Wu, X., Ren, C., Joshi, T., Vuong, T., Xu, D., and Nguyen, H. T. (2010). SNP discovery by high-throughput sequencing in soybean. *BMC Genomics* 11, 469. doi:10.1186/1471-2164-11-469

Xie, W., Feng, Q., Yu, H., Huang, X., Zhao, Q., Xing, Y., Yu, S., Han, B., and Zhang, Q. (2010). Parent-independent genotyping for constructing an ultrahigh-density linkage map based on population sequencing. *Proc. Natl. Acad. Sci. U.S.A.* 107, 10578–10583.

You, F. M., Huo, N., Deal, K. R., Gu, Y. Q., Luo, M. C., Mcguire, P. E., Dvorak, J., and Anderson, O. D. (2011). Annotation-based genome-wide SNP discovery in the large and complex *Aegilops tauschii* genome using next-generation sequencing without a reference genome sequence. *BMC Genomics* 12, 59.

Young, M. D., Wakefield, M. J., Smyth, G. K., and Oshlack, A. (2010). Gene ontology analysis for RNA-seq: accounting for selection bias. *Genome Biol.* 11, R14.

Zhang, G., Guo, G., Hu, X., Zhang, Y., Li, Q., Li, R., Zhuang, R., Lu, Z., He, Z., Fang, X., Chen, L., Tian, W., Tao, Y., Kristiansen, K., Zhang, X., Li, S., Yang, H., and Wang, J. (2010). Deep RNA sequencing at single base-pair resolution reveals high complexity of the rice transcriptome. *Genome Res.* 20, 646–654.

INDEX

www.ingramcontent.com/pod-product-compliance
Lightning Source LLC
LaVergne TN
LVHW062222200926
848484LV00006B/1455